THE ECONOMICS OF PERSISTENT INNOVATION: AN EVOLUTIONARY VIEW

Economics of Science, Technology and Innovation
VOLUME 31

Series Editors
Cristiano Antonelli, *University of Torino, Italy*
Bo Carlsson, *Case Western Reserve University, U.S.A.*

Editorial Board:
Steven Klepper, *Carnegie Mellon University, U.S.A.*
Richard Langlois, *University of Connecticut, U.S.A.*
J.S. Metcalfe, *University of Manchester, U.K.*
David Mowery, *University of California, Berkeley, U.S.A.*
Pascal Petit, *CEPREMAP, France*
Luc Soete, *Maastricht University, The Netherlands*

The titles published in this series are listed at the end of this volume.
Economics of Science, Technology and Innovation

THE ECONOMICS OF PERSISTENT INNOVATION: AN EVOLUTIONARY VIEW

edited by

William R. Latham
University of Delaware

and

Christian Le Bas
Institut des Sciences de l'Homme

 Springer

ISBN-10: 1-4899-7362-7   ISBN 0-387-29245-4 (eBook)
ISBN-13: 978-1-4899-7362-7

9 8 7 6 5 4 3 2 1

springeronline.com

CONTENTS

LIST OF FIGURES

LIST OF TABLES

LIST OF CONTRIBUTORS

Alexandre Cabagnols
Assistant Professor, Clermont-Ferrand University and University of Lyon 2
Laboratoire d'économie de la firme et des insitutions,
14, avenue Berthelot , F-69363 LYON cédex 07
alexandre.cabagnols@univ-bpclermont.fr

Nilotpal Das
Statistician, Futures, LLC
409 Executive Drive, Langhorne, Pennsylvania, 19047 USA
n_das167@yahoo.com

Claudine Gay
Professor, Catholic University of Lyon and University of Lyon 2
Laboratoire d'économie de la firme et des insitutions,
14, avenue Berthelot , F-69363 LYON cédex 07
cgay@univ-lyon2.fr

William Latham
Associate Professor of Economics, University of Delaware
Department of Economics, Newark, Delaware 19711 USA
latham@udel.edu

Christian Le Bas
Professor of Economics, University of Lyon 2
Laboratoire d'économie de la firme et des insitutions,
14, avenue Berthelot , F-69363 LYON cédex 07
lebas@univ-lyon2.fr

James Mulligan
Professor of Economics, University of Delaware
Department of Economics, Newark, Delaware 19711 USA
mulligaj@lerner.udel.edu

Karim Touach
Researcher, University of Lyon 2
Laboratoire d'économie de la firme et des insitutions,
14, avenue Berthelot , F-69363 LYON cédex 07
Karim.Touach@univ-lyon2.fr

PREFACE AND ACKNOWLEDGEMENTS

Christian Le Bas

The idea of studying the persistence of firm innovative behavior emerged gradually over the period during which I directed the Economy and Applied Statistics Laboratory[1] at the University Lumière Lyon 2 in the middle 1990s. Definitive studies were carried out from the very beginning by enthusiastic and productive young researchers. First Alexandre Cabagnols dealt with persistence as a substantial part of his PhD thesis. Then Claudine Gay identified the crucial roles of consistent and persistent inventors in the collective process of knowledge growth. Subsequently, with the assistance of Karim Touach, Claudine and I have undertaken new empirical research on persistent inventors as relatively unknown figures in inventive structures. This nascent enterprise received an intellectual and logistical boost several years ago with the collaborations of William Latham and James Mulligan, my colleagues from University of Delaware (USA). The completion of this book owes much to their efforts and especially to the productive links I have forged with William Latham, the co-editor, my co-author and my friend. We are indebted to the highly successful cooperative exchange program between the University of Delaware and the University Lumière Lyon 2 which encouraged the development of the professional contacts that led to our collaboration. While it is hopeless to try to acknowledge all of those whose comments, remarks and criticisms have enriched our work, I must make special mention of Paolo Saviotti whose salient comments contributed so vitally to Chapter 7.

This project was accomplished with precious, and mostly persistent, support from a variety of sources including the funding for two different research groups by the University Lumière Lyon 2 (first the Economy and Applied Statistics Laboratory and then the Economics of the Firm and Institutions Laboratory[2] and grants from the French National Center for Scientific Research[3] program, "Economic Issues in Innovation."[4] I cannot forget the Catholic University of Lyon's program (GEMO-ESDES) which provided me with the time for writing a significant part of this book. Finally I wish to thank the editors and publishers of the collection at Springer for accepting the risk of publishing the manuscript.

I dedicate this book first, to the memory of Ehud Zuscovitch, with whom I shared both a passion for evolutionary economics and a friendship and, second, to Keith Pavitt, a true intellectual leader in the field of the economics of innovation.

---

[1] Laboratoire d'Économie et de Statistiques Appliquées (LESA)

[2] Laboratoire d'Economie de la Firme et des Institutions

[3] Centre National de la Recherche Scientifique

[4] Les enjeux économiques de l'innovation."

INTRODUCTION

William Latham
Christian Le Bas

Persistence of firm innovative behavior became an important topic in applied industrial organization with the publication of the seminal empirical work of P. Geroski and his colleagues (1997). Evidence that firms innovate persistently has led previous studies to focus on the determinants of innovation persistence and on its heterogeneity across industries, technologies and countries. The aims of this book are: (1) to illumine the scale and scope of the phenomenon of persistence in innovation, and (2) to account for the principal factors that explain why some firms innovates persistently and others do not.

Because this book deals intensively and extensively with the subject of firm innovation persistence, which is not, as yet, a well-known term, we need to provide a nontrivial definition of it that encompasses the full range topics we want to address and aids our understanding of how they are related to each other. We begin with a careful identification of "innovation." Our first definition is drawn from K. Pavitt (2003), "innovation processes involve the exploration and exploitation of opportunities for a new or improved product, process or service, based either on an advance in technical practice or a change in market demand, or a combination of the two." While this definition is clear, and conforms well to both our empirical and theoretical perspectives, some elaboration may help to clarify the concept. For example, in empirical quantitative studies, including those of this book, the choice of a measurable indicator of innovation brings additional nuances to the definition. Pavitt (2003) argues that a simple improvement of an existing product ought to be included as an innovation. This means that innovation is not necessarily "radical" or "architectural" but, often, very often indeed, merely "incremental."[1] However, while innovation occurs at the level of the individual firm it is not a "firm-in-isolation" phenomenon: what is new for only a single firm within an industry, cannot be considered to be an innovation for the industry or the economy as a whole.

Our view of the appropriate definition of innovation has implications for the economic analysis of innovation. Nelson and Winter (1982) distinguished three strategies for firm technological development: innovation, imitation and "no change." It is clear to us that a large number of previous studies addressing innovation persistence have combined innovation and imitation as a single strategy as the alternative to "no change" (see, for instance, Saviotti, 2003). Our own definition of innovation is obviously similar to the one implicitly accepted by the community of researchers in economic studies of innovation persistence.

The second part of the term that identifies our subject of interest in this book, "innovative persistence" is persistence. We need also to define what we mean by persistence. Fortunately common usage accords well with our usage of the term in this case. By persistence we mean, in part, "continuing to occur over time." We also mean, as will be further explained by Das and Mulligan in Chapter 6, "continuing to occur over space." Generalizing the concept we will recognize as persistent any behavior initiated at one point and subsequently observed at related points. The nature of the relationship may be purely temporal, temporal and spatial, or across other spaces in which firms operate including industrial and technological space.

Until relatively recently little empirical evidence on the innovative persistence phenomenon had been assembled and, in addition, no systematic theoretical framework has yet been suggested for understanding persistence in innovation. This book aims to fulfill this dual gap. We present new empirical evidence that the contributing authors have assembled and suggest a coherent theoretical framework. We will present arguments in favor of an evolutionary competence/capability approach to the phenomenon of persistence in innovation. The authors who support such an evolutionary theory of innovation, either explicitly or tacitly, utililize a vision of the firm rooted in behavioral theory (Metcalfe, 1995). In behavioral theory firms have the capacity for learning and exhibit adaptive behaviors. In general, firms do not maximize any objective function in particular, because economic information is difficult to gather and to analyze. In the technological arena, there is an additional reason for optimizing conduct not to be the dominant mode: creativity, and especially technological creativity, is fundamentally an uncertain process. Creativity and innovation are connected to diversity across firms as well. Each firm uses its own particular visions and routines to explore the technological and economic opportunities it meets, and exploits them in its own particular ways. Economists have attempted for some time to find regularities within the innovation process in order to understand this diversity among firms. Pavitt's (1984) well-known taxonomy of sectoral technological trajectories is among the best attempts, though it is still tentative, for explaining this diversity. It is based on the simple idea that firms from different sectors develop innovation differently. The rates and the directions of technical change experienced by a firm depend on three firm characteristics:

the sources and the nature of the firm's technological opportunities,

the nature of the firm's technological requirements, and

the possibilities for innovating firms to appropriate the benefits of their innovating activities.

The last feature is required if firms are to have incentives to invest resources in research and other innovative activities such as design (Dosi et al. (1990, pp 90ff)). Pavitt identifies four general sectoral technological trajectories:

(a) science-based sectors (electronics, chemicals),

(b) scale-intensive sectors (automobiles, consumer durables),

(c) specialized-supplier sectors (machinery, instruments), and

(d) supplier-dominated sectors (private services, traditional manufacturing), in which firms buy innovation through their capital goods.

Table 1 illustrates some salient sectoral characteristics of the technological trajectories in terms of

(i) firms' sources of new knowledge,

(ii) firms' price and/or performance sensitivity,

(iii) firms' means of protecting innovations,

(iv) firms' sources of process technology, and

(v) types of innovation (product versus process).

Regarding the last of these, Von Tunzelman (1995) asserts that the measure of an organization's technological effectiveness is its success in transforming knowledge about technologies (processes) into knowledge about products.

A continuing theme of this book is that the study of innovative persistence must explicitly consider the specific features of the relevant innovative trajectories. Thus in Chapter 3 Alexandre Cabagnols shows how the idea of sectoral innovative trajectories can be applied and provides insightful commentary. Nilotpal Das and James G. Mulligan in Chapter 6 explicitly focus their analysis on the adoption of process innovation of a sort that is typical in a "supplier-dominated" trajectory.

Table 1: Innovation characteristics of four sectoral technological trajectories

Sectoral Characteristics	Innovative Trajectory Sectors			
	Science-based	Scale-intensive	Specialized-supplier	Supplier-dominated
Source of New Knowledge	R&D and Public Science	Product Engin-eering	Design and Development	Suppliers and Large Users
Price or Performance Sensitive	Mixed	Price Sensitive	Performance Sensitive	Price Sensitive
Means of Appropriation	Patents	Trade secrets	Design know-how	Non-technical
Sources of Technology	In-house + Suppliers	In-house	In-house + Customers	Suppliers
Innovation Type: Product Versus Process	Mixed	Process	Product	Process

Adapted from Dosi et al. (1990)

The book is structured as follows. Chapter 1 elaborates important basic themes and definitions and presents a brief survey of the previous literature. In Chapter 2 we provide a first empirical analysis of the principal determinants of innovation persistence following the lines opened by of Geroski et al. (1997). The data are for French industrial firms patenting in the US. The results emphasize the importance of firm size and the existence of a minimum threshold of innovative activity. In Chapter 3 Alexandre Cabagnols uses several French innovation surveys to evaluate firm competences that promote innovation (entry) and those that maintain the firm in a dynamic of persistence in innovation. He shows it may be that the two sets of competences are not identical.

In Chapter 4 a fascinating new perspective on innovative persistence is presented: the role of persistent individual inventors. Persistent inventors are individual inventors for whom we find large numbers of French, German, British and Japanese patents in the National Bureau of Economic Research's U.S. patent database. The presence of such inventors, who are consistently inventing, is used to explain and predict some mechanisms underlying general patent activity. In Chapter 5 Alexandre Cabagnols explores the impact of the level of technological accumulation of French and UK firms on their ability to persist in innovation over longer periods of time (1969-84). Cabagnols estimates a Cox model in which the stock of technological knowledge enters as a time varying covariate. In particular he observes that the impact of past patenting activity on the development of subsequent innovations decreases very quickly in both countries (he finds a depreciation rate of 60%). Both the French and the UK samples lead to qualitatively similar results. In Chapter 6 Nilotpal Das and James G. Mulligan analyze evidence concerning persistence in the adoption of innovations by firms that do not create innovations themselves. Until recently the economic literature has ignored persistence in the adoption of subsequent vintages of technologies by adopting firms. The Chapter contains original empirical work extending recent results to account for the persistence of adoption across vintages of ski-lift technology. They find for example, that firms that adopted the earliest vintages were most likely to adopt newer versions of the technology. This is counter to the possibility that firms might delay adoption in anticipation of a newer version appearing in the future. The authors argue that persistence in this case is due to the firm's incentive to differentiate the quality of its service from that of its competitors. Chapter 7 sets out an evolutionary approach to persistence in innovation. We first identify the foundation of evolutionary principles upon which a non-formal analysis of innovation persistence can be built. Then we propose a more formal model incorporating important features of the evolutionary tradition. The model is shown to be capable of accounting for a number of real-world observations and of facilitating some interesting insights regarding the nature of innovative persistence. In the Chapter 8 we discuss the main findings set out in this book, suggest new future research agenda and draw some implications in terms of public policy.

ENDNOTES

[1] For additional definitions of innovation see Tushman and Anderson, 2004

REFERENCES

Dosi, G., Pavitt, K., Soete L., (1990), The Economics of Technological Change and International Trade, Brighton: Wheatsheaf, and New York: New York University Press.

Geroski, P., Van Reenen, J., Walters C. F., (1997), "How persistently do firms innovate?" Research Policy, vol. 26, pp. 33-48.

Metcalfe, J. S., (1995), "The Economic Foundations of Technology Policy: Equilibrium and Evolutionary Perspectives," in: P. Stoneman, (ed.), Handbook of the Economics of Innovation and Technological Change, Blackwell, pp. 409-512.

Nelson, R. R., Winter, S. G., (1982), An Evolutionary Theory of Economic Change, London: The Belknap Press of Harvard University Press.

Pavitt, K., (1984), "Sectoral patterns of technological change: Towards a taxonomy and theory," Research Policy, vol. 13, pp. 343-374.

Pavitt, K., (2003), "The Process of Innovation," SPRU Electronic Working Paper Series, n° 89, August.

Tushman, M., Anderson, P., (2004), Managing Strategic Innovation and Change, A Collection of Readings, 2nd, Oxford: Oxford University Press

Chapter 1

PERSISTENCE IN INNOVATION: DEFINITIONS AND CURRENT DEVELOPMENT OF THE FIELD

Christian Le Bas, *University of Lyon 2*
William Latham, *University of Delaware*

1. INTRODUCTION

The objectives of this chapter are to provide the reader with a more comprehensive understanding of the concept of persistence in innovation, to distinguish its roles in each of two distinct analytical frameworks, to describe its current status as a research area, and finally to identify its principal determinants. First we elaborate our definition of the subject. Then we discuss the significance of persistence in innovation both in the standard neoclassical approach of industrial organization and in a Schumpeterian model of industrial dynamics. Next we provide a survey of the principal empirical literature on the subject. The final section of the chapter provides an overview of what we believe are the main factors determining persistence in innovation.

2. PERSISTENCE IN INNOVATION: DEFINITIONS

We consider two possible definitions of firm persistence in innovation. In the first definition, the persistence of innovative behavior is identified if a firm that has innovated during a given time period also innovates in the following time period. A very simple figure can be used to describe this process. Assume two time periods (t and t+1) and two possible actions for the firm under observation in each time period: to innovate or not to innovate. The four possible cases are illustrated in Figure 1.

Firm's Action in Period t+1	Firm's Action in Period t	
	Innovate	Do Not Innovate
Innovate	Case 1	Case 3
Do Not Innovate	Case 2	Case 4

Figure 1. Firm Innovative Behavior in Two Periods: Four Cases

In Case 1 the firm innovates continuously or persistently. The firm innovates sporadically in Case 2 and Case 3 and the firm does not innovate at all during the time periods considered in Case 4. The sporadic patterns of innovation described by Cases 2 and 3 are not equally likely. The process described in Case 3 is less likely than the process in Case 2. A firm which does not innovate (in time t) is more likely to be a firm which does not invest in R&D or in other technological activities (engineering, design, etc.), or it is a firm which has not built routines, organizational structures, and competencies for succeeding in its technological projects. Such a firm has less chance to survive to the next period of time, and, if it survives, it is not likely to have enough resources for investing in technological activities. As a result, the probability of such a firm innovating in t+1 is very low. Case 2 applies to firms that experience "success and failure in innovation". The concept described in Figure 1 can be extended to many time periods. A firm that innovates during only one period of time is an "occasional" innovator or a "single-shot" innovator, and a firm that innovates from time to time is a "sporadic" innovator"[1].

Next consider a second, stricter, definition of persistence in innovation. We consider that the particular technological field in which an innovation is realized (and in which a patent may be granted) matters. For two periods we still have our four cases as in Figure 1 but now we will explicitly consider innovation only in technological field *j*. Now a firm is considered to innovate "persistently" only if it produces an innovation in the **same field** in the two periods of time (See Figure 2, Case 5). If it does not, we get a technological exit or death (Case 6) or a technological natality (Case 7). Extending the biological metaphor, Case 8 represents "technological sterility". If a firm innovates only in technological field *j* in period *t* and then in innovates again in *t+1*, but not in field *j* but in another one, say field *i*, then, with our stricter definition, the firm does not innovate persistently with respect to field *j*. Its innovation in field *i* will be recorded as a technological natality in that field. Malerba and Orsenigo (1999) were the first to point out the importance of technological entry or exit. They distinguished between "real" entrants (exiters) and "lateral" entrants (exiters): "real" entrants are those firms that did not innovate previously in any technological class and "real" exiters are firms that ceased to innovate in any technological class. In contrast "lateral" entrants and exiters are those firms that innovated in the past in a different technology. It is clear that the significance of the distinction between real and lateral entrants depends strongly on the level of aggregation (the number and breadth of technological classes) at which the analysis is carried out. At the most highly-aggregated, macro level, the two definitions are equivalent.

3. PERSISTENCE IN INNOVATION: THE STANDARD APPROACH OF INDUSTRIAL ORGANIZATION VERSUS SCHUMPETERIAN MODELS OF INDUSTRIAL DYNAMICS

The standard approach of industrial organization emphasizes the incentive structure for explaining each of these processes. The seminal paper of Arrow (1962) showed that a when a monopolist innovates he basically just replaces himself as the monopolist in his market. This is called the "replacement effect" in the theory of innovation (Tirole 1988). Because of this replacement effect, a monopolist gains less from innovating than does a competitive firm which converts a formerly competitive market into a monopoly through innovation. If we observe a firm which has innovated and now has monopoly power, we will predict that such a firm has lower incentives for innovating again than would a competitive firm. Thus, this approach, based on the "replacement effect," predicts less persistence in innovation. This would explain Case 3 in Figure 1. This model may be very too restrictive in its assumptions for predicting general results. Gilbert and Newberry (1982) have studied a more interesting case: a monopoly threatened by a potential entrant. In this case it is possible to demonstrate, under very general assumptions, that, because entry will reduce its profit, the monopolist's incentive to remain a monopolist is greater than the entrant's incentive to become a duopolist. There is clearly an asymmetric effect between the insider whose new innovation will destroy rents generated from prior innovations and the outsider, who loses nothing from innovating. This could produce Case 1 or Case 2 when the monopolist has innovated far in the past. Similarly, in the framework of patent races, Gallini (1992) has pointed out that the threat posed by losers deciding to continue their research programs due to their sunk costs creates an incentive for the patent race winner to improve his own invention. This is clearly an argument in favor of the persistence of innovation.

Another situation which may favor innovation persistence arises when "pre-emption" is the firm's optimal strategy (Tirole, 1988). In this situation a firm may "buy" an innovation to prevent its competitor from using or implementing it. If the firm was innovating itself prior to the purchase, it might be able to exploit its own and the purchased innovations jointly. Many results from R&D rivalry models such as these depend critically upon unverifiable assumptions concerning the distributions of information, the identities of the decision variables and the sequences of changes (Cohen, Levin 1989). Nevertheless, the main limitation that we see with these models is that they work in only one direction: from market structure to innovative (or non-innovative) behavior. Rarely (excepted in the static frame of economic games) are market structure and innovative behavior jointly determined. In contrast in evolutionary models the focus is on the

> dynamic process by which firm behavior patterns and market outcomes are jointly determined *over time*....Through the joint action of search and selection, the firms evolve over time, with the condition of the industry in each period bearing the seeds of its condition in the following period. (Nelson and Winter 1982, p. 1819).

The co-evolution of market structure and innovative behavior is at the core of evolutionary analysis.

Firm's Action in Period t+1	Firm's Action in Period t	
	Innovate in Field *j*	Do Not Innovate in Field *j*
Innovate in Field *j*	Case 5	Case 7
Do Not Innovate in Field *j*	Case 6	Case 8

Figure 2. Firm Innovative Behavior in Two Periods in Technological Field j: Four Cases

In the last decade evolutionary theory has been improved by many scholars. Technological natality and firm innovative persistence have been considered as two important aspects of the creation of technology at the core of the Schumpeterian view of economic development. On one hand, natality is associated with the entrepreneurial technological regime (Winter 1984) or with the Schumpeter Mark 1 model (Acs and Audretsch 1991, Malerba and Orsenigo 1995, 1996) in which the firms are small, competition is strong, and turbulence (entry and exit) is high. On the other hand, firm persistence in innovation is related to a routinized technological regime (Winter 1984) or to the "Schumpeter Mark 2 model" (Malerba and Orsenigo 1993, 1996) in which the firms are (very) large, oligopolistic competition is stable, and turbulence is weak. These two models illustrate Schumpeterian dynamics: the first describes the "creative destruction" of firms and activities; the second describes the "creative accumulation" of knowledge within the large, persistent firms which develop dynamic competences (Teece and Pisano, 1994). The evolution of firms within these models interacts strongly with the industry life-cycle. In the first part of the life-cycle the entrepreneurial technological regime favors entry into the industry. For this reason innovation occurs within small firms, but mortality is high and selection severe (Winter 1984). In the mature phase of the life-cycle a routinized technological regime predominates. Innovation occurs in the R&D departments of large oligopolistic firms which have a longer economic life than small firms in the entrepreneurial regime. We would expect that a small firm will innovate over a shorter period of time than a large firm. The fact that some firms innovate over a short period of time and others over a long period of time should be considered as normal in an industrial environment characterized by diversity and variety.

4. PRIOR EMPIRICAL ANALYSES OF PERSISTENCE IN INNOVATION.

In the following literature survey we use our first, broader definition of firm innovative persistence which does not distinguish the technological field or class of innovations. Geroski *et al.* (1997) use two sets of data: patent applications and direct innovation data concerning UK firms. These data show that very few firms are persistently innovative. Geroski *et al.* find that the number of patents granted at the beginning of an innovative period (what we will call a "spell") is a good and strong predicator of the length of such spells. When they use innovation data, they include employment in their regressions as a measure of the size of the firms. They find that larger firms have longer innovation spells. But the relationship between the size and length of spells seems to be "highly non-linear." When they use patent data, however, they are not able to test the effects of the size of the firm on the length of innovation spells for their whole sample. Estimates using a smaller sample of firms predict that there is a positive but weak effect of pre-spell firm size (measured by turnover). Finally, it appears to be clear that the volume of patenting or innovation activity prior to the spell considered explains spell length better than firm size. These results are not surprising. They tend to emphasize that the main factor behind the persistence of innovation is the size of "innovation activity" (measured, for instance, by the volume of R&D expenditures) more than the size of "economic activity." Regarding the mechanisms that would explain this dynamic, Geroski *et al.* propose a combination of "learning effects" in the production of innovation and positive feedbacks between accumulation of knowledge and the production of innovation All things considered, the production of innovation would be subject to dynamic economies of scale. Because Geroski *et al.* provided a feasible quantitative methodology for estimating econometric relationships in this area, their study has been considered as a benchmark for further studies.

Duguet and Monjon (2002) use the European Community Innovation Survey (CIS) 1 and CIS 2. They merge these two data sets and obtain a sample of 808 firms operating in the manufacturing sectors in the 1986-1996 of time period. They retain a broader definition of innovation. Moreover they study persistence in innovation behavior in a shorter period of time. Their first descriptive analysis shows that innovation persistence is strong. They subdivide the manufacturing sector into industries in which technological opportunities are high those in which opportunities are low. In high-tech industries the proportion of persistent innovators is significantly higher: the proportion of persistent innovators in high-tech industries is three times what it is in low-tech industries. Duguet and Monjon also observe persistence of non-innovators. The proportion of firms that do not innovate at all (persistent non-innovation) is also high: 35% of the firms that do not innovate in one period of time also did not innovate in the previous ones (Case 4 in Figure 1). Duguet and Monjon's results also show that the proportion of firms we consider as occasional innovators or a "single-shot" innovators (or as "sporadic" innovators when they innovate from time to time) is low, constituting only 20% to 30% of the sample. The fact there is a persistence of non-innovating behavior is a little surprising in the sense that both conventional and evolutionary economics predict that such firms will not survive in the long run. However, on should bear in mind that the CIS, from which the data are drawn, records managers' declarations about innovation and not direct observation on it. We know there can be a gap between an individual manager's subjective view of what constitutes a technological innovation and the observation of an economist or the inference drawn from objective data. Some managers only classify revolutionary (or major) changes as "technological innovation" and consider their firm's incremental (or minor) improvements, which are, in fact innovations, as not significant. In this way some firms will identify themselves as non-innovative while, in reality, they definitively are.

A number of studies have used econometric methods to obtain quantitative results. These studies have shown that the probability of innovating in any period is strongly explained by the fact the firm has innovated in the two previous periods with the effect weaker for the earlier period. This result is robust to the introduction of variables such as the size of the firm and industry sector characteristics. Other studies have proposed that the lagged innovation variables in these estimations are not really capturing the effects of a process of learning in innovative activities, but are simply proxies for the effects of prior R&D. That is, R&D carried out in past explains the current level of innovation. With respect to this problem these studies emphasize the extreme importance of formal R&D realized within the firm. The regressions suggest that, when firms invest persistently in formal R&D programs, innovative persistence results: the persistence in innovation comes from persistence of R&D. On the other hand when the formalization of the R&D function is less pronounced, the coefficient related to past innovation is more important. The key point is that underlying the process of persistence in innovation is another persistence process which affects R&D activities. In fact it is the same process as the one in innovation activities: the first is seen from the point of view of inputs (R&D), the second from the point of view of output (innovation).

Lhuillery's (1996) analysis of French industrial firms obtained interesting results regarding persistence. He uses European Community Innovation Surveys (CIS) to attempt to explain why firms cease to innovate. He finds that the firm's industrial sector matters. Firms from the traditional sector cease innovating more frequently. He also finds that, given the sector, the size of the firm is a good predictor f persistence. Small-and-medium-sized firms seem to have more difficulty sustaining persistent innovation. Other factors are also found to play a role, including the type of innovation. For example, the more radical an innovation is, the more probable it is that the firm will stay on an innovative trajectory. Acquisition of knowledge (including R&D results) by the firm is another factor favoring innovative persistence. There is also a link between the proportion of the firm's revenues derived from innovating products and the probability of innovating in the current period. Finally, Lhuillery reports the existence of a sectoral demand effect: the stronger are demand pressures, the higher is the probability that a firm innovates persistently. This result provides support for economists who think a large part of the innovation process is demand-driven (Le Bas and von Tunzelman, 2004). The differences between process and product innovation in the context of persistence are also addressed, but in more general terms (see Chapter 3 of this volume in which this topics is explored as well).

Malerba *et al.* (1997) use European patent data for six countries over the 1978-91 period. They study technological entry and exit and the process of turbulence. They confirm results from Geroski *et al.* (1997): a large fraction of new innovators are occasional. In a Schumpeterian framework, persistence in innovation is related to technological cumulativity, learning-to-learn and new research opportunities. The longer innovation spells are dependent on accumulated knowledge. In association with industrial heterogeneity these phenomena produce concentration and stability in the ranking of innovators and low turnover in the population of innovators, all characteristics of a "routinized" regime (Winter, 1984). As far as is concerned Malerba *et al.* (1997) find significant country-effects for turbulence: stable in Germany and Japan, very high in Italy, and medium in France and the UK. Another empirical study from Malerba and his colleagues (Malerba and Orsenigo, 1999) using the same European patents data base yields interesting insights. They observe that a large fraction of new innovators ceases to innovate soon after entry into the industry; consequently, they are occasional innovators. Malerba and Orsenigo point out that most entrants have only one patent. It follows that, if these firms represent a large fraction of the whole population of innovators, they hold a small share of total patenting activity. Only a small proportion of new entrant-innovators continue to innovate. These entrant-innovator firms grow in size and also in terms of innovative activity. They survive as persistent innovators. Unfortunately, Malerba and Orsenigo do not have enough information construct profiles of these firms in terms of sectors, technological fields, countries and other characteristics identified above.

Cefis (1999) uses a panel of 82 UK firms repeatedly observed over the 1978-91 period to jointly analyze the innovative activities of firms that are occasional and systematic innovators and their profitability. Her results suggest that firms that are systematic innovators earn profits above the average and have a strong incentive to keep innovating and earning profits above the average.

Le Bas and Négassi (2002) examine the effects of persistence in innovation on sectoral innovative intensity and on international technological specialization. They use European patent data for 1996-98 (plus some patent data information for 1990-92) for France, Germany, and the UK. Their sample contains all firms that applied for five or more patents during the period under observation. They obtained additional economic information for a sample of 2130 firms in France (14233 patents), 2673 in Germany (23214 patents) and 2425 in the United-Kingdom (11187 patents). Sectoral innovative intensity is measured by the number of patents applied for in a technological field divided by an indicator of the sector size, the number of firms which apply for at least one patent in this field. This indicator may be biased by small firms. To evaluate the relative technological strengths (or weaknesses) of the different countries, they use an index of revealed technological advantage (RTA)[2] Two main variables characterize the Schumpeterian patterns of innovation: technological natality (the newly created firms which patent for the first time) and firm innovative persistence. All of their regressions are log-log and the method of estimation is GLS. They introduce dummies to control for sectoral characteristics. The results show that the structural variables have a strong and significant impact on sectoral innovative performance. In particular, the size of the sector explains a significant part of the sectoral innovativeness. The influence of demand conditions (the Schmooklerian influence) is weaker and less significant when firm innovative persistence is introduced in the regressions. The sign of an industrial concentration variable changes when natality and firm innovative persistence are permutated. This could be due to the fact that firm innovative persistence and industrial concentration are negatively correlated. The results show that natality and firm innovative persistence have significant and positive impacts. It is possible that these key variables, which describe the characteristics of the firm's technological environment, also capture some aspect of market structure and industrial competition. Finally there is evidence that a sector is more innovative when sectoral demand is strong, the size of the sector is large, and both the technological natality and firm innovative persistence are significantly high.

Next Le Bas and Négassi estimate four models of the index of technological specialization (RTA). Their findings indicate clearly that the sectoral innovative intensity determines the RTA. This means that when firms from one country innovate intensively in a field, this nation is strong in this technological field. This condition is necessary, but not always sufficient. Le Bas and Négassi also examine the impacts of three other variables: demand conditions, size of the sector and technological concentration. It appears that they have no significant influence on RTA. Persistence is equally a good and strong predicator of RTA. This confirms the point of view expressed in favor of the impact of large firms. Technological natality has no impact on RTA. The dummies, which are supposed to take into account the country characteristics in the models, are significant only when the index for persistence is introduced. Finally the RTA is high in technological areas where the firms are persistently innovative (and probably large), and the industrial concentration is high. The technological natality (due greatly to small new firms) has an impact on sectoral innovative intensity, but not on international technological specialization once we control for technological sectoral characteristics.

The main conclusion we can draw is that innovative firm persistence is an important phenomenon which affects technological innovativeness across sectors (the sectoral intensity of innovation) and countries (international technological specialization). So the two aspects of the Schumpeterian dynamics (creative destruction and creative accumulation) co-determine innovativeness at the sector level. But only creative technological accumulation (a large-firm dynamic capability) has an important impact on international technological specialization (measured by the RTA).

A recent study from Brandt (2004) shows that in new industries, at the beginning of their industry life-cycles, new firms (new entrants) are important for spurring innovation. The converse is true for mature industries, marked by lower entry rates and lower levels of expenditures on formal R&. In the first group of industries, innovation is correlated to the rate of entry; in the second group, productivity is correlated to innovation persistence.

Athrye and Edwards (2002) study of innovation persistence is of interest for two reasons: First, they examine the factors explaining persistent innovation among small and medium enterprises (SME) in the UK. The innovation continuity of SMEs is not often studied, largely because much of the data, such as that on patenting and innovation surveys, has little information on SMEs' innovating activities (except in the high-tech sectors). Second, they utilize survey data collected by the Centre for Business Research at the University of Cambridge. Athrye and Edwards estimate a bivariate probit model for two regions. Their results are in accord with those of previous studies for variables such as size and type of innovation. They find that persistence in innovation seems to be influenced by environmental factors, but that these factors are not statistically significant in distinguishing one region from the other. Recently economic analysis has increasingly emphasized the importance of local or regional clustering on firm innovative performance. New studies are needed to measure the importance of the regional (if not local) environment on innovation persistence.

Finally it appears that many diverse indicators have been used in the investigation of persistence in innovation. Each gives a slightly different picture of this phenomenon. For example, using a sample of Dutch firms, van Leuven (2002) has shown that innovation performance over time is lower when persistence is measured on the output side (innovation) than when measured on the input side (R&D). But it seems very clear, too, that persistence in innovation exists and will be an important phenomenon to understand whatever the indicator used.

Based on the findings of the empirical studies reviewed in the preceding paragraphs concerning the persistence in innovation, we can conclude with confidence that:

1) Firm size is an important determinant of innovative activity size. In fact, a minimum threshold size for total revenues (turnover) appears to be required for the firm to be able to fund permanent (persistent) R&D activity and to have the possibility of innovating. But this relationship between innovative persistence is certainly not linear (Pavitt et al., 1987) and does not have the same form in all industrial sectors. This may explain why innovation spell length is better explained by the number of patents at the beginning of the spell as a proxy for the size of innovative activity.

2) The size of innovative activity influences the degree of technological variety.

3) There is a minimum threshold size of innovative activity necessary to become a persistent innovator (Geroski et al. (1997) and Le Bas et al. (2003)). This explains why small patentees patent in a short period of time, and why heavy (consistent) patentees are persistent innovators.

In others words, only consistent innovators become persistent innovators (see Malerba and Orsenigo 1999).

4) The type of industry matters. In high-tech industries the scale of innovative persistence is higher than in low-tech industries. Similarly mature industries have more persistence in innovation than new industries.

5) The type of innovation is important (Lhuillery, 1996).

6) There is a strong relationship between persistence in innovative behavior and persistence of above average profits.

4. THE DETERMINANTS OF INNOVATION PERSISTENCE.

In this section we review the main explanations of innovation persistence. A good starting point for understanding innovation persistence is to consider demand and supply factors for technology. Von Tunzelman (1995) considered two types of influences on technological activities: demand-pull influences (*e.g.,* the demand for technology or innovation) and supply-push influences (*e.g.,* the supply of technology or innovation). The demand for innovation focuses on the incentives, pressures, or signals for modifying existing technology that originate on the demand side of the market. The supply of innovation focuses on the costs of generating new knowledge (*i.e.,* the productivity or efficiency of the innovation process). There are several notable characteristics of the supply for technology:

- An increase in the supply of technology is the outcome of "technological accumulation" due to the growth of the firm's knowledge through learning. This growth reduces the costs of knowledge production whether in the form of learning-by-doing or formal scientific learning.
- Inter-firm linkages, either horizontal or vertical, facilitate the transfers of knowledge and thereby increase the supply of technology. This type of transfer is likely to occur whatever the degree of sectoral competitiveness. There are many forms in which such transfers can occur, such as learning by interacting, technological collaboration, or establishment of technological districts. Sometimes the establishment of linkages to facilitate transfers, such as the establishment of a technological district, requires action by a third party, an "agent of technology."

It is clear that technology is not determined solely by a single force. It is not a case of demand for technology *versus* supply of technology. Practically always the two types of influence act together but with different intensities.[3]

A major purpose of a firm's research, engineering and design activities is to balance the demand and supply sides of technology to produce the best level and type of technology for the firm. Our analysis of technology in the form of innovative persistence must therefore also maintain a balance between the more traditional, cost-based supply determinants of knowledge production and the more recent emphasis on demand-side factors (provided "the demand influences and their pathways are properly understood," as Von Tunzelman has stipulated. (Von Tunzelman 1995, p.420). The balancing between the two types of forces depends on a number of factors, including, among others:

- Sectoral technological characteristics including the type of technological trajectory followed. Among the types of technological trajectories identified by Pavitt (1984), there are some for which the impact of demand for innovation is stronger than the impact of supply side factors.
- The segment of the technology-cycle from which the dominant technologies are being derived.
- The national system of innovation policies.

Of course there are many factors accounting for variations in the rates of innovation in different sectors. For example, variations in market structures, including variations in degrees of industry concentration, will produce: (1) variations in incentives to invest in R&D, (2) variations in demand pressures, 3) variations in levels of technological opportunity, and (4) variations in regimes for protecting innovations. The notion of firm innovation persistence seems to be different because its main determinants are not located in the firm's external economic environment in the way that the above factors are, but in its own internal structure. At first it seems that a virtuous dynamic internal to the firm plays the unique role. However, in fact there is always interplay between factors external to the firm and internal factors in determining persistence in innovation.

Geroski at al. (1997) proposes the existence of "dynamic economics of scale" in the production of innovation and in the process of innovation persistence as follows: the volume of innovation produced by a firm at a given period of time has an effect on the "quantity" of innovation realized later. It may be that learning in innovation activity can explain this phenomenon. This notion accords well with the well-known view that R&D has two faces: innovation and learning (Cohen and Levinthal, 1989). Learning here is a capacity to innovate later. There are many views of the way in which learning-by-doing works in research activity. It could be that a major innovation in a firm in one time period stimulates a sequence of minor innovations in the next time period. The empirical work of Bell and Bourgeois (1999) suggest such a process is working in the energy sector: the large firms which produce a major innovation will improve it a few years later through the means of cumulative incremental innovations. In this case, we easily understand why there is persistence in innovative behavior. Due to its technological leadership (which may be as derived from nothing more than the existence of lead times to implement innovations) the innovator is first in designing and implementing new relevant technological improvements at lower economic costs. It might be this process is in action when some minor innovations stimulate the sequence as well. These examples illustrate particularly well the technology push approach.

The theory of induced innovation offers an interesting explanation for the persistence of innovative behavior as well. In this theory bottlenecks develop due to a gap between demand and supply of specific products. These bottlenecks tend to increase prices and thus provide additional incentives to firms for to engage in research on new materials or new products which will establish a new balance between demand and supply. For example, Mowery and Rosenberg (1979) note that innovation is often triggered in responses to demands for the satisfaction of certain classes of needs. Many researchers have found that market demand has a decisive governing influence on technological innovation. Some studies emphasize the importance of "demand-pull" influences and minimize the potential effectiveness of supply-push policies. So the role of demand has sometimes been overextended and misrepresented. It is now generally acknowledged both the knowledge base of science and technology and the structure of demand play large roles in innovation as an *interactive or coupling process.*

There is another, more macroeconomic, economic theory of innovation: the so-called "theory of biased-induced innovation." In this theory market forces push individuals firms to seek out inventions which have specific factor-saving biases. Generally this process is a response to changes in relative scarcities at the macroeconomic level, the macro-demand being larger than the macro-supply (see the seminal contributions of Fellner 1961, Kennedy 1965, and Samuelson 1965). In this sense innovations are induced by demand. This type of analysis was integrated into growth theory in the 1970s.

We believe that research turns money into knowledge and that innovation turns knowledge, including knowledge coming from markets, into money. Knowledge is clearly at the core of innovation and thus must also be at the core of the innovation persistence process as well. We will show later in this volume in Chapters 2 and 7 how this understanding leads us to emphasize the role of a firm's "dynamic capabilities." But this does not mean that the mere existence of "dynamic economies of scale" in the production of innovation necessarily results in a pure virtuous knowledge cycle in which each phase inevitably produces the next. Innovation turns knowledge into products that must be accepted by the market in order for the cycle to continue. In this way market demand is the filter which effects economic selection. This explains why demand and supply factors interact so strongly. What we must explain is why, for a great majority of firms, this interaction does not work well. In other words, we need to explain why firms fail frequently and why only a few firms innovate persistently. The same model which explains innovation persistence must explain sporadic (irregular) innovations as well.

ENDNOTES

[1] Many innovations having their origin in "demand-pull" have "supply for technology" determina[1] Of course every firm either innovates or does not in every time period, however the time periods are defined. The shorter the time periods, the lower will be the frequencies of observed time periods with innovations. Our terms are meant to evoke the idea that the relative frequency of innovation varies across firms regardless of the time periods under consideration. The premise of this paper is that is important to understand the nature and determinants of this frequency distribution whether firms exhibits persistence in innovation or not.

[2] The RTA index varies around unity, such that values greater than one indicate that the country is relatively strong in the technology considered as compared to other countries in the same technological field, while values less than one indicate a relative weakness.

[3] Many innovations having their origin in "demand-pull" have "supply for technology" determinants as well (Von Tunzelman 1995).

REFERENCES

Brandt N. (2004), "Business Dynamics, Regulation and Performance," STI Working Paper 2004/3.

ACS, Z.J., and Audretsch, D.B., eds. (1991), *Innovation and Technological Change*, London: Harvester Wheatsheaf.

Athreye, S. and Edwards S. (2003), "Persistence in Innovation: Evidence from Data on UK SMEs," Paper presented at a conference in honor of K. Pavitt, University of Sussex, 13-15 November.

Arrow, K.J. (1962), "Economic Welfare and the Allocation of Resources for Invention", in : R. R. Nelson, ed., The Rate and Direction of Inventive Activity: Economic and Social Factors), Princeton, NJ: National Bureau of Economic Research and Princeton University Press.

Bell, F., and Bourgeois, B. (1999), "Innovation Direction and Persistence within an Industry: The Refining Processes Case", Paper presented at the European meetings of the Applied Evolutionary Economics Association, 7-9 June 1999, Grenoble.

Bresnahan, T., and Greenstein, S. (1999), "Technological Competition and the Structure of the Computer Industry," Journal of Industrial Economics, March 1999, 47, pp. 1-40.

Cabagnols, A. (2000), "Les Déterminants des Types de Comportements Innovants et de leur Persistance: Analyse Evolutionniste et Etude Econométrique," Unpublished Ph. D. Dissertation, University of Lyon 2.

Cabagnols, A. (2003), "Technological Learning and Firm Persistence in Innovation: A France/UK Comparison Based on a Cox Model of Duration", Working paper, Centre Walras, University of Lyon 2.

Cefis, H. (1999), "Persistence in Profitability and in Innovative Activities", Paper presented at the European meetings of the Applied Evolutionary Economics Association, 7-9 June 1999, Grenoble.

Cefis, H. (2003), "Is there persistence in innovative activities?" International Journal of Industrial Organisation, 21 (4), pp. 489-515.

Dosi, G. (2000), Innovation, Organization and Economic Dynamics, Northhampton, MA: Edward Elgar.

Duguet, E., and Monjon, S. (2002) "Les Fondements Microéconomiques de la Persistance de l'Innovation", Revue Économique, 55 (3), pp. 625-636.

Gallini, N.T. (1992), "Patent Policy and Costly Imitation," RAND Journal of Economics, 23 (1), pp. 52-63.

Geroski, P., Van Reenen, J., and Walters, C. F. (1997), "How Persistently Do Firms Innovate?" Research Policy, 26, pp. 33-48.

Gilbert, R. and Newberry, D. (1982), "Pre-emptive Patenting and the Persistence of Monopoly," American Economic Review, 72 (3), pp. 514-526.

Iosso, T.R. (1993), "Industry Evolution with a Sequence of Technologies and Heterogenous Ability. A Model of Creatrice Destruction," Journal of Economic Behavior and Organization, 21, pp. 109-129.

Le Bas, C., Cabagnols, A., and Gay, C. (2003), "An Evolutionary View on Persistence in Innovation: An Empirical Application of Duration Model, " in : P. Saviotti, ed., Applied Evolutionary Economics, Northhampton, MA : Edward Elgar.

Le Bas, C. and Négassi, S. (2002), Les structures des activités d'innovation en France et comparaison avec celles des principaux partenaires commerciaux. Convention d'étude n° 19/2000 COMMISSARIAT GENERAL DU PLAN. Final report Nov. 2002.

Le Bas, C. and von Tunzelman N. (2004), "Croissance industrielle, demande, et activités technologiques," Revue d'Économie industrielle, Mars.

Lhuillery, S. (1996), "L'innovation dans l'Industrie Manufacturière Française: Une Revue des Résultats de l'Enquête Communautaire sur l'Innovation," in Innovation, Brevets et Stratégies Technologiques, Paris: Organisation for Economic Co-operation and Development.

Malerba, F., and Orsenigo L. (1995), "Schumpeterian Patterns of Innovation", Cambridge Journal of Economics, 19, (1), pp. 47-65.

Malerba, F., Orsenigo, L., and Peretto, P. (1997), "Persistence of Innovative Activities, Sectoral Patterns of Innovation and International Technological Specialization", International Journal of Industrial Organization, 15, pp. 801-826.

Malerba, F., and Orsenigo, L. (1999), "Technology Entry, Exit and Survival : An Empirical Analysis of Patent Data," Research Policy, 28 (6), pp. 643-660.

Mansfield, E. (1962), "Entry, Gibrat's Law, Innovation and the Growth of the Firm", American Economic Review, 52 (5), pp. 1023-1051.

Mansfield, E. (1986), "Patents and Innovation: An Empirical Study", Management Science, 32 (2), pp. 173-181.

Marris, R. (1964), "A Model of the Managerial Enterprise," Quarterly Journal of Economics, 77 (2), pp. 185-209.

Metcalfe, J.S. (1993), "Some Lamarkian Themes in the Theory of Growth and Economic Selection," Revue Internationale de Systémique, 7 (5), pp. 487-504.

Metcalfe, J.S., and Gibbons, M. (1986), "Technological Variety and the Process of Competition", Économie Appliquée, 3, pp. 493-520.

Mueller, D.C. (1997), "First-Mover Advantages and Path Dependence", International Journal of Industrial Organization, 15 (6), pp. 827-850.

Saviotti, P. (1996), Technology Evolution, Variety and the Economy, Cheltenham, UK and Brookfield, VT: Edward Elgar Publishing, Limited.

Teece, J., Pisano, G. (1994), "The Dynamic Capabilities of Firms: An Introduction," Industrial and Corporate Change, 3, pp. 537-556.

Tirole, J. (1988), The Theory of industrial Organization. Cambridge, MA : MIT Press.

Van Leuven, G. (2002), "Linking Innovation to Productivity Growth Two Waves of the Community Survey," OECD STI Working Paper 2002/8.

Von Tunzelman, G. N. (1995), Technology and Industrial Progress. Edward Elgar

Chapter 2

DETERMINANTS OF PERSISTENCE IN INNOVATION: A STUDY OF FRENCH PATENTING*

Alexandre Cabagnols, *Clermont-Ferrand University and University of Lyon 2*
Claudine Gay, *Clermont-Ferrand University and University of Lyon 2*
Christian Le Bas, *University of Lyon 2*

1. INTRODUCTION

In this chapter we report the results of an empirical analysis using a methodology very similar to the one implemented by Geroski et al. (1997). Our aims are simple: 1) to provide the reader with a preliminary understanding of some of the difficulties encountered in conducting empirical analysis of innovation persistence, and 2) to identify the variables likely to be the most important in explaining this phenomenon. Our empirical analysis uses a sample of 3 347 French industrial firms that made patent applications to the US Patent office from 1969 to 1985 and had a total of 22 000 patents granted. We estimate the coefficients of variables which explain the duration of innovation (the length of the innovation period) by using an econometric model of duration.[1] In the following section we describe our data set after first reviewing the arguments in favor of patents as an indicator of innovative activity.

2. THE DATA

2.1 PATENTS AS INDICATORS OF INNOVATIVE ACTIVITY

Following the seminal work of J. Schmookler (1966), patent statistics are now widely used as indicators of innovative activity and as proxies for innovation in econometric estimations.[2] However the use of patents has been

*A first draft of this chapter was published as chapitre 8 in P. Saviotti ed. Applied Evolutionary Economics. Edward Elgar. 2003. We thank Edward Elgar who gives the permissison to reuse and modify the first draft.

criticized because this indicator has limits as a measure of technical change: patents are only one means of protecting innovation, and their importance as protection against imitation varies amongst sectors. Patents are not a good measure of radical innovations associated with new paradigms such as software and bio-technology. Moreover, raw counts of patent grants may fluctuate over time simply because of constraints on the number of patent officers and budgetary allocations of the patent office (Z. Griliches, 1990) rather than fluctuations in the technology levels.

In spite of these acknowledged drawbacks, patents offer many advantages. Patents play a significant role at every stage of the innovation process (B. Basberg, 1982). Patents are also an indicator of technological competence. Here we define technological competence as the ability to transform technological and market opportunities into an activity or knowledge that leads to industrial production (B. Carlsson and G. Eliasson, 1994). As the codified and publicly available outcome of cognitive and tacit problem-solving processes, a patent is the result of a process of accumulation and the production of technological knowledge. In spite of all associated problems, patent statistics are generally acknowledged to provide a rough proxy for innovation.[3]

2.2   USING   PATENTS   GRANTED   TO   FOREIGN PATENTEES BY THE US PATENT OFFICE.

An important limitation of the use of patents as technological indicators in cross-country studies is related to the differences among various countries' laws on property rights. To solve this problem, and following K. Pavitt and L. Soete (1980), it is now common to use American patents held by patentees from other countries rather than to use the national patents of the other countries. This is particularly important in economic studies dealing with technological accumulation. In fact, due to a uniform institutional framework for all patents regardless of country of origin within the U.S. patent system, the comparison of patenting by patentees from different countries is possible. Such comparisons would not be possible using patents from different national patent offices because of the significant differences in national patent systems. Use of data from the U.S. patent system also allows us to compare our work with the results of Geroski et al. (1997). The U.S. patent system provides a wealth of information because of the sheer size of the American market for technology which results in huge numbers of patents even for patentees from individual foreign countries within the U.S. system. In addition the system's allocation procedures have remained constant over time and it has been relatively accessible in electronic form. Foreign-owned patents in the U.S. system may be even better indicators of innovation by the foreign counties than their own domestic patents within the foreign countries themselves if it can be assumed that it is not worthwhile (in terms of effort and expense) to obtain U.S. patents for low value innovations. (D. Archibugi, 1988, D. Archibugi and J. Michie, 1995, B. Basberg 1983, and P. Patel, 1995).

Eventually the European patent system will offer many of the same advantages now held by the U.S. system. In fact, since the creation of the "Observatoire des Sciences et des Techniques" (OST) in France, numerous studies have examined European patents. However, because the European system of patents only started in 1978 and did not become fully operative until 1985, it cannot yet be used to assess the capacity to innovate over a long period of time.

There are three particular limitations in using patents as indicators of technological innovation for studying persistence:

Patents do not account or all the innovations which are patentable, in particular those for which secrecy is the best way of protecting them.[4] This is crucial for some industrial sectors and our analysis risks biased results because of it. We attempt to deal with this, in part, through the use of complementary information from an innovation survey (see A. Cabagnols, 2000 and S. Lhuillery, 1996).

Patent data do not account for possible weak connections between innovative activities and patent applications. A firm may invest in R&D continuously but receive patents only sporadically, say one or a group of related patents only every other year. Such a pattern is especially likely for small firms. We have tried to restrict the significance of this problem in our data by adopting the convention that single year breaks in patenting are not sufficient to identify a discontinuity in innovative activity.

Patent data may inadequately reflect industrial sector characteristics. It is likely that each industrial sector has its own general way of innovating, the "sectoral technological trajectory" and, thus, its own dynamics of persistence in innovation.

3. STATISTICAL ANALYSIS OF PATENTING SPELLS.

In order to explore the reasons why some industrial firms innovate over a very short period (2 or 3 years) and others over long time periods (more than 10 years), we begin by exploring our patent data with the use of some simple statistics. Table 1 provides statistics for a panel of 3 347 French industrial firms, observed over the period 1969 to 1985, which have been granted patents by United States Patent and Trademark Office at some time during this period. First, it is clear that we have in our panel only a small part of the 25 000 French firms which have more than 20 workers. Second, we have in our panel not only French firms but some subsidiaries of foreign firms which are located in France. Our panel encompasses all the French industrial firms which have been granted patents in the US. Unlike Geroski et al. (1997) it is not a balanced panel. Each firm can patent over one or several spells[5]. Finally, we retain the convention that two spells must be separated by a two-year time period at least. To put it simply, a one year break without being granted a patent is not sufficient to consider that there has been a break in the process of innovation. By our convention, a two-year break shows that innovation is not continuous or persistent. There are several reasons for this. Patents are simply an indicator of innovative activity, not a precise measure of the timing of the innovation process. In particular they cannot inform us about the time period in which innovation has exactly occurred. This is even more the case for the granting of patents than patent applications. This argument is stronger for French firms which were not accustomed to patenting abroad (particularly in the 1960s and 1970s) and which might sometimes have decided to cluster their patenting projects over a number of months before making an application, perhaps to take advantage of scale economies in legal costs.

We are able to conduct a statistical analysis of the total number of spells and of the maximum length of spells (the longest spell for each firm). In fact a firm is able to innovate during several periods, each separated from the next by the two years required to identify separate instances over the period under observation. In such cases the same firm will appear several times when we study spells, as we do in this paper. Another option is to retain in the sample of spells only one spell, the longest spell (or the maximum spell) for a firm. The time distribution of patenting spells is given in Table1. The main feature is that of 3902 spells, 71%, are very short, one-year spells. As the length increases from 2 to 15 years the number of spells decreases monotonically. There is a small spike for the longest spell (16 years). We basically found the same features of the distribution of spells in our data as was found in the data studied by P. Geroski et al. (1997). As regards the maximum length of spells, the frequency distribution is very similar. Figure 1 shows clearly that a large proportion of spells is located on the left for the shortest spells (less than 4 years).

Table 1. Distribution of patenting spell lengths and maximum spell lengths by firm for 3347 firms, 1969-85

Length of Spell (years)	Number of Spells			Maximum Spell Length		
	No.	%	Cumul %	No.	%	Cumul %
1	2855	73.17	73.17	2394	71.53	71.53
2	336	8.61	81.78	288	8.6	80.13
3	262	6.71	88.49	230	6.87	87,00
4	98	2.51	91,00	90	2.69	89.69
5	65	1.67	92.67	63	1.88	91.57
6	58	1.49	94.16	55	1.64	93.22
7	52	1.33	95.49	51	1.52	94.74
8	27	0.69	96.18	27	0.81	95.55
9	19	0.49	96.67	19	0.57	96.12
10	20	0.51	97.18	20	0.6	96.71
11	17	0.44	97.62	17	0.51	97.22
12	14	0.36	97.98	14	0.42	97.64
13	14	0.36	98.33	14	0.42	98.06
14	13	0.33	98.67	13	0.39	98.45
15	10	0.26	98.92	10	0.3	98.75
16	42	1.08	100.00	42	1.25	100.00
Total	3902	100.00		3347	100.00	

Table 2 gives information on other features of patenting activity. The distribution of total patents per spell has a mean of 5.65, a median of one patent, and a standard deviation of 35.23.

Table 2. Descriptive statistics for 3902 patenting spells in 3347 firms, 1969-1985

Variables	Std Dev	Mean	Med	Sum	Min	Max	Lower Qrtile	Upper Qrtile	Variance	Skewness	Kurtosis
NB	35.23	5.65	1	22044	1	1329	1	2	1240.93	21.17	617.8
DSPELL	2.57	2.02	1		1	16	1	2	6.61	3.57	13.64
VT		1.65	1		1	22	1	1	3.26	4.75	29.68
VTDP		1.15	1		1	10	1	1	0.36	6.74	63.66
DSPMAX	2.64	1.83	1		1	16	1	1	6.99	3.42	12.83
NBDP		1.29	1	5017	1	30	1	1	1.36	11.99	208.21

NB : number of patents during a spell

DSPELL : length of spell over which firm patent (years)

VT : technological variety (the number of different technological fields in which a firm patents during a spell)
VTDP : initial technological variety (the number of different technological fields in which a firm patents in the first year of a spell)
DSPMAX : length of the maximum spell over which a firm patents (years)

NBDP : initial number of patents (the first year of a spell)

The distribution of spells has a mean of 2.202 years and a standard deviation of 2.570. For each of the 3347 firms in the panel we are able to determine the longest spell of innovative activity or the spell maximum (DSPMAX). The distribution of this variable has a mean of 1.83 years and a standard deviation of 2.64. The distribution of total patents per firm (not shown in Table 2) has a mean of 6.59, substantially less than the mean of 10.8 in the panel of British firms used by Geroski et al., a median of one patent compared with Geroski et al.'s 2, and a standard deviation of 38.02, also less than the 79.98 of Geroski et al. In others words, British firms patent more with a greater dispersion. This could be due to two complementary reasons: British firms are more innovative and/or are more accustomed to making patent application in the US. Figure 2 shows that a large majority of patents is located on the right that is to say, for the longest spells.

4.   DESCRIPTION OF VARIABLES IN THE ANALYSIS

Two variables are very important in our model: the number of different technical fields in which a firm patents during one spell or technological variety (VT), which has a maximum value of 34 in our sample, and the technical field in which a firm patents most during one spell (CTmax). Firstly, let us consider the indicator of technological variety VT. There is a strong and very significant correlation between VT and the spell length (R2 = 0.64). But correlation is not same as causality. Here, as it is often the case for the phenomenon of duration, we suspect that the causality runs in both directions: the longer the spell, the higher the probability of patenting in another technical field and, at the same time, the greater a firm's multi-technological knowledge, the greater are its dynamic capabilities,[6] and the longer its innovation spells will be (these data are, in fact, different for each firm). For this reason, we will take into account another variable, VTDP, which measures technological variety at the beginning of the firm's spell (these data are also different for each firm). The mean of the VTDP distribution is smaller than the mean of VT. The two distributions are strongly correlated: VTDP explains 50% of VT variance. Because VTDP cannot be determined by the spell length, we intend to use it as a variable which explains the persistency of innovation behavior. In our framework, we argue that such a variable (VTDP) is a good measure of the firm's capacity to maintain efficient technological production over several years. We have therefore chosen it as a proxy for the dynamic capabilities of firms.

CTmax, that is to say the technical field in which a firm patents most during the spell, defined at the firm level, is the technical class in which the firm patents the most. In our framework it is another term for firm core-technological competency. There is an obvious interest in using it, too, as a proxy for technological opportunity. The empirical literature states that technical advance is easier or less costly in some industries than in others (W. Cohen, 1995). It is particularly relevant to use this notion for assessing why some firms are able to innovate over a longer period of time than others. Given the difficulty of constructing technological opportunity measures for samples encompassing numerous industries, we treated technological opportunity as a dummy variable. For convenience and to facilitate statistical treatment, CTmax is actually represented by six qualitative dummy variables for the six major technological fields: CTmaxA for Electrical and Electronic, CTmaxB for. Instruments, CTmaxC for Chemicals and Drugs, CTmaxD for Metallurgical and Chemical Processes, CTmax E for Mechanical and Transport, and CTmaxF for Consumer Goods and Construciton. In order to assess the relative weight of each group of competences we calculated the

CTmax associated with each spell. In our sample we obtained the following results: CTmaxA=10.56%,   CTmaxB=8.94%,   CTmaxC=16.76%%, CTmaxD=17.99%, CTmaxE=39.80%, CTmaxF=5.95%. This distribution confirms:

The continuing and widespread (or "less and less" neglected) importance of improvements in mechanical technologies (P. Patel and K. Pavitt, 1994)

The importance of patenting in drugs and chemicals industries, for which a patent is the best mean of protection.

When core technological competences are dominated by chemical-drug technologies, the spells have a mean duration of 2.4205 years. The corresponding figures are 2.4077 years for electrical-electronic technologies, 2.0431 years for mechanical/transport, 1.722 years for metallurgical and chemical processes (that is to say lower than the mean of the panel of 2.0169 years), 1.6963 for instruments, and 1.3836 for consumer goods/construction technologies. This variable explains 39% of the variance of spell length. The results are highly significant. Firm core technological competences tend to partly explain innovation spell lengths. These findings are very similar to those in the empirical literature (see particularly W. Cohen, 1995).

The foregoing can be summarized in a taxonomy of regimes of patenting behavior (see Table 3). Four types of behavior are defined:

Single patentees patent over short spells (3 years maximum), producing few patents. This group comprises 71% of the sample (64% in P. Geroski, et al., 1997) and produced 1.146 patents per firm.

Heavy patentees patent over long spells (13 years maximum) producing many patents. This group represents 2.2% of the sample (2 % in Geroski, et al., 1997) and produced 136.9 patents per firm.

Moderate patentees take out between 2 and 10 patents over one spell only. This group represents 12.2% of the sample (this group is not identified by Geroski, et al., 1997) and produced 14.1 patents per firm.

Sporadic patentees patent over several spells. This group represents 13.9% of the sample (34% of the sample for P. Geroski, et al., 1997, but the two studies do not use the same definition of the spell lengths) and produced 7.02 patents per firm.

Table 3.   A taxonomy of regimes of patenting behavior

	Definition	Number of firms %	Number of patents %
Single Patentees	1. One short spell of 3 years maximum 2. Few patents (2 maximum)	2397 (71.62%)	748 (12.47%)
Heavy Patentees	1. One long spell of 13 years minimum 2. Many patents (10 minimum)	75 (2.24%)	10268 (46.58%)
Moderate Patentees	1.One spell 2. Between 2 and 10 patents	408 (12.19%)	5749 (26.08%)
Sporadic Patentees	1. More than one spell	467 (13.95%)	3279 (14.87%)
Total		3347 (100.00%)	22044 (100.00%)

5.   NON-PARAMETRIC ESTIMATION USING THE KAPLAN-MEIER ESTIMATOR

In order to assess the effects of the variables we first use a non-parametric estimation. Figures 1a and 1b shows empirical survival rates and associated standard errors for the panel as a whole and Figure 2 for spells starting with 1, 2, 3, 4, 5 and more patents at the beginning of the spell. The percentage of spells that lasts at least as long as T years is computed using the Kaplan-Meier estimator (see Greene, 1997). Survival rates for spells with one patent in their year are about 28%, for those with 2 patents 51.5%, for those with 3 patents 78%, for those with 4 patents 88%, and for those with 5 patents or more 90%.

These results are very similar to those found by Geroski et al. (1997). A firm which begins a spell with only one patent at time t is less likely to patent in t+1 than a firm that commences a spell with 2 patents and so on. When we consider the survival rates for spells which last 2 years, we find that their chance for surviving a third year is respectively 18.5%, 43%, 69%, 83%, and 90%. Figures 1a (for the panel as the whole) and 1b (for each group of spells) illustrate graphically these trends. Firms with one patent have only a 13 % probability of enjoying a patenting spell of 5 years. This probability increases with the number of patents at the beginning of the spell. When a firm begins to patent 5 or more times, the probability of surviving as an innovator is 82%.

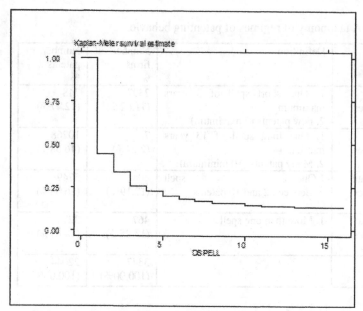

Figure 1. Kaplan-Meier survival estimates.

Geroski and his colleagues did not obtain such scores, but the 20-years period they studied is somewhat longer than ours and covered a different set of years. The higher the level, in terms of patents at the beginning of the patenting period, the greater the chance of surviving as a patentee. The degree of relative disadvantage declines as the initial level of patenting increases, but more rapidly than in Geroski et al. (1997). So it seems there is something that resembles "dynamic scale economies" in the patenting activity.

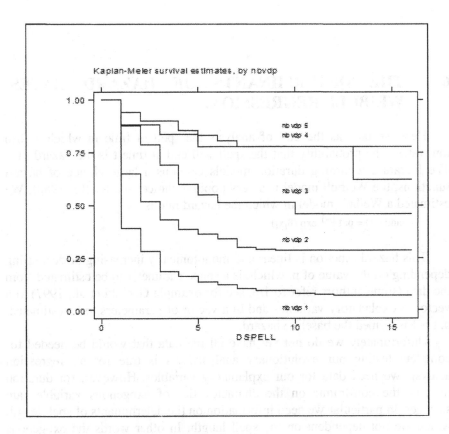

Figure 2. Kaplan-Meier survival estimates, by nbvdp

We also have calculated (but not reported in this chapter) the survival rates for the spells which are associated with each type of technological competence (CTmax). A firm which begins a spell associated with electrical-electronic competences is more likely to patent the second year than a firm that commences a spell associated with chemical-drugs competences, which is more likely to patent a second year than a firm that commences a spell associated with metallurgical-chemical processes competences, and so on. These results confirm that the effects of technological opportunity are differentiated along each group of competences.

6.   THE DETERMINANTS OF HAZARD RATES: WEIBULL REGRESSION.

Here we take as the unit of analysis the spell of time in which a firm innovates. The probability that the spell will end at time t is the hazard rate. The literature regarding duration models contains a large choice of hazard functions; the Weibull model is a very popular choice (see Kieffer 1988). We estimated a Weibull model in which the hazard rate is:

$$\text{landa } (t) = p.\, t^{\,p-1} \exp(ß_i x_i)$$

This hazard function is linear and monotonically increasing or decreasing depending on the value of p, which is a shape parameter to be estimated from the data (some authors refer to 1/p, see for example Geroski et al., 1997) xi a vector of explanatory variables and ßi a vector of parameters to be estimated. p. t p-1 is termed the baseline hazard.

Unfortunately we do not yet have all the data that would be needed for complete testing our evolutionary analysis. As is true for all regression analysis we need data for our explanatory variables. However, for duration models, the constraints on the characteristics of exogenous variables are stronger. In particular we need information on the determinants of spell length which are not dependent on the spell length. In other words the exogenous variables must be assumed not to change from the beginning of the spell to the "failure time." For example, the total number of patents granted during a spell is clearly proportional to the spell duration (the longer the spell the greater the number of patents granted). For these reasons we are only able to use the estimation variables which describe the firms' structural characteristics or initial levels of variables (at the beginning of the spell). More generally it is difficult to find information concerning firms' economic characteristics over such a long period of time. We finally decided to insert as explanatory variables technological variety at the beginning of the spell (VTDP), a good predicator of the dynamic capabilities of the firm, and the initial level of patenting (nbdp) as a set of categorical dummy variables with five categories for initial levels of 1, 2, 3, 4, and 5 or more. This set of dummy variables defines the capacity of the firm to maintain innovation activity over a long period of time. In the regression nbdp5 is the omitted category. The technical field in which firms patent the most, with the six categories defined above) is a proxy for technological opportunities. We use this set of dummy variables to control for the effect of VTDP on patenting spells.

Regression 1 in Table 4 shows clearly that technological variety at the beginning of the spell is a variable which explains spell length. In regression 2 we add CTMax as a dummy variable. In regression 3 we introduce other dummy variables (nbdp) without any other explanatory variables. Our main findings are:

The estimates are very significant (except for nbdp4 in regression 3).

For the 3 regressions we find: $1/p = 0.85$. This result is very close to those of Geroski et al. (1997) who found $1/p = 0.5$ with patent data, and $1/p = 0.7$ with innovation data. The findings are highly inconsistent with the hypothesis of negative duration dependence.

The variable nbdp clearly has a significant impact (except for nbdp4). This confirms the Kaplan-Meier survival analysis of the preceding section.

The effects of the dummies variables concerning the types of technological competences mastered by the firm, a proxy for technological opportunities, conform to our expectations based on the preceding statistical analysis. Chemical-drugs technological competences (CTmaxC) and electrical-electronic competences (CTmaxA) have the most important fixed-effect (in the regression CTmaxF is the dummy omitted), in regression 3 we add nbdp as a dummy variable.

When we add the explanatory variable VTDP in regression 3, none of the dummies are significant. Here we suspect that there is a correlation between VTDP and nbdp. The link is obvious: the greater the number of patents at the beginning of the spell, the more firms are likely to patent in different technical fields (among the 34). The link between VTDP and the dummy CTmax is less clear. It is probably due to the fact that large firms are more represented in chemicals, drugs and electronics than in other fields of technological competence. Firm size positively influences the degree of diversity of firms' technological activities (P. Patel, K. Pavitt 1996): large firms are able to patent in several fields. These remarks mean there is probably a firm-size effect which is a determinant of spell length through pre-spell patenting and technological variety. This presumption in favor of a size effect is discussed by Geroski et al. (1997). It seems that firm size has necessarily an effect on pre-spell patenting (see F.M. Scherer 1965 for an analysis of the relation between firm size and patenting). Finally our opinion is that the relationship between firm size and spell length is complex and should be re-examined seriously with other data.

One remarkable finding is the confirmation of the hypothesis of a positive duration dependence $(1/p>1)$. In other words, the likelihood of failure at time t, conditional upon its value up to time t, is increasing in t. But this general law, which governs the hazard function, is coupled to an "economic" law which stipulates that there are threshold effects concerning patenting at the beginning of the spell. Technological variety, a proxy for dynamic capabilities of the firms, plays the same role. On the basis of a case study of the refining industry, that is, on the basis of only one intra-industry analysis, F. Bel and B. Bourgeois (1999) confirm that there is a strong relationship between large technical portfolios and persistent innovation.

Table 4. Results of Weibull regressions of patenting spells

	1	2	3
Const.	0.2241603*** (0.0716743)	0.5260703*** (0.0991457)	-3.806565*** (0.3271031)
VTDP	-1.211365*** (0.0650397)	-1.190869*** (0.065051)	
nbdp1			3.220226*** (0.3178645)
nbdp2			2.205539*** (0.3224657)
nbdp3			1.330229*** (0.3528568)
nbdp4			0.4419265 (0.5164873)
Ctmaxa		-0.4772388*** (0.0966871)	-0.5060203*** (0.0967425)
Ctmaxb		-0.2524355** (0.0987095)	-0.2491801* (0.987398)
Ctmaxc		-0.5307958*** (0.0890269)	-0.5863441*** (0.0892373)
Ctmaxd		-0.1861146* (0.0872036)	-0.2637319*** (0.0872634)
Ctmaxe		-0.3445846*** (0.811755)	-0.3583056*** (0.812273)
ln p	0.1438304*** (0.0125765)	0.155501*** (0.0126167)	0.1733986 *** (0.0126043)
1/p	0.8660346	0.8559862	0.8408024
Pseudo-R	0.08002	0.08559	0.094615
Khi-2 (k-1)	803.55***	857.37***	948.37 ***
log likelihood	-4610.1	-4583.27	-4537.77

Method of estimation : maximum likelihood
Dependent variables : log of time until failure es
Number of observation : 3902
Standard errors given in parentheses
For the Khi-2 model k is the number of regressors
For the Khi-2 model k is the number of regressors
** = significant at the 1/100 threshold
* = significant at the 5/100 threshold

7.   CONCLUSIONS.

In this chapter, we have developed an evolutionary competence based approach for explaining the persistence of French firm innovative behavior. Until now this phenomenon has not really been analyzed. In this book we basically are concerned by innovation persistence, not the persistence of patenting. Up to this point we have considered patents as unbiased indicators of innovation. At this point of our analysis it is time to take into account the likely gaps between patent and innovation. Patenting is one among others means for appropriating the economic returns to innovation. The relative importance of patenting differs greatly across technologies and consequently across industries (Cohen et al., 1997). For this reason the propensity to patent an invention is not the same across industries.[7] In this spirit we must interpret with caution our results regarding the effects of the core competences of the firm on its own innovation persistence. In the same vein, the propensity for patenting may eventually shifts over time with new regulations concerning the patent system. New types of technological knowledge, which formerly did not enter directly in the legal patent framework, may begin to do so at some time. For these two reasons we must be cautious in the interpretation of results. This being said, our results appear to be robust.

We mostly confirm the main results from Geroski and al. (1997), using US patent data for 1969-1985. Few firms are persistently innovative. Only 42 firms of a sample of 3902 innovate continuously over the 17 year-period of time studied. 73 per cent innovate over a very short one-year period of time. 14% are sporadic patentees. This last case is very interesting and should generate more empirical work. Concerning the results of our Weibull estimation, the main findings are positive duration dependence and threshold effects concerning patenting at the beginning of the spell. We find new significant results at the firm level as regards the impact of technological variety and the dynamic capabilities of firms or technological opportunities over spell length. F. Bel and B. Bourgeois (1999) have suggested a very interesting explanation of persistence in innovative behavior. If we have understood the above analysis, a firm (probably large) which realizes a major innovation will improve it few years later though the means of cumulative incremental innovations. In this case, we can easily understand why there is persistence in innovative behavior. This type of process is another way of emphasizing firm "dynamic capabilities." There is a good reason for carrying out empirical case studies of persistence at the industry level as F. Bel and B. Bourgeois have done with many details on the innovations.

Returning to the issues addressed by Geroski regarding the effects of firm size and (firm) innovation activity size as predicators the length of innovation spells, our opinion is:

Firm size is necessary a determinant of innovation activity size since a minimum threshold of turnover appears required for the firm be able to fund permanent (persistent) R&D activity and possibly to innovate. But this relation is certainly not linear (evidence is given by Pavitt et al. 1987), and does not have the same form in all industrial sectors. It is probably why innovation spell length is better explained by the number of patents at the beginning of a spell, a proxy for the size of innovation activity

The size of innovation activity influences the degree of technological variety (captured here by VTDP), too.

It could be there is a minimum threshold of innovation activity necessary for becoming a persistent innovator. Evidence is given by Geroski et al. (1987) and by our own estimates regarding the effects of the number of patent in the pre-spell period. This would explain, on the one hand, why small patentees patent in general over short periods of time and, on the other hand, why heavy (consistent) patentees are persistent innovators. In others words, only consistent innovators could become persistent innovators.

In the following chapter the relevance and the coherence of findings will be questioned at two levels. The use of a Weibull model rests on particular assumptions; we need to test its generality. The same is true for the empirical results. In analyzing data from other countries, we will gain understanding regarding its generality as well.

ENDNOTES

[1] For more about these models see N. Kieffer (1988) and C. Gourieroux (1989).

[2] See, in particular, B. Crepon and E. Duguet (1994), Z. Griliches (1990), A. Pakes, (1985), F. M. Scherer (1986), and L. Soete (1981).

[3] Patents are only a rough proxy because so much innovation is not patented as documented by A. Arundel and I. Kabla (1999). Using the PACE/CMU results they found that a patent application is made for only approximately a third of all innovations.

[4] See the preceding footnote for more on this issue.

[5] Recall that a spell is the length of time (number of years elapsed) from the first year of patenting until the year when we observe that the firm has ceased innovating. In our data set all firms in the panel have at least one patent in 1969 so that is the initial of the first spell for all firms.

[6] The concept of "dynamic capabilities" points in the same direction as the concept of "core competence" (see the Introduction in Dosi et al.,eds., 2002).

[7] For example, patenting may be a product of a firm's lobbying to make its strategic position-better as well. Some firms can apply for patents for strategic purely reasons. This could explain why some firms patent persistently without necessarily innovating persistently.

REFERENCES

Archibugi, D. (1988), "The inter-industry distribution of technological capabilities. A case study in the application of the Italian patenting in the U.S.A.", Technovation, 7, pp. 259-274.

Archibugi, D., Michie, J. (1995), "The Globalisation of Technology : a new taxonomy", Cambridge Journal of Economics, Vol. 19(1), pp. 121-140.

Arundel, A., Kabla, I. (1998), "What percentage of innovations are patented? Empirical estimates for European firms", Research Policy, 27, pp. 127-141.

Basberg, B. (1982), "Technological change in the Norwegian whaling industry - A case study in the use of patent-statistics as a technology indicator", Research Policy, 11, pp. 163-171.

Basberg, B. (1983), "Foreign patenting in the U.S. as a technology indicator", Research Policy, 12, pp. 227-237.

Bel, F., Bourgeois, B. (1999), "Innovation direction and persistence within an industry: the refining processes case." Communication to the European Meeting on Applied Evolutionary Economics, 7-9 June 1999, Grenoble.

Carlsson, B., Eliasson, G. (1994), "The Nature and Importance of Economic Competence", Industrial and Corporate Change, 3(3), pp. 687-711.

Crepon, B, Duguet, E. (1994), "Research and Development, competition and innovation: what the patent data show", Centre de Mathematiques Economiques, CNRS URA 924 12

Cohen, W. (1995), "Empirical Studies of Innovative Activity", in P. Stoneman (ed.), Handbook The Economics of Innovation and Technical Change, Basil Blackwell, Oxford.

Dosi et al. Eds (2002) The Nature and Dynamics of Organisational Capabilities. Oxford University Press.

Geroski, P. (1994), Market Structure, Corporate Performance and Innovative Activity, Oxford University Press, Oxford.

Geroski, P., Van Reenen, J., Walters, C. F. (1997), "How persistently do firms innovate?" Research Policy, 26, pp. 33-48.

Gilbert, R., Newberry, D. (1982), "Pre-emptive Patenting and the Persistence of Monopoly", American Economic Review, 72, pp. 514-526.

Gourieroux G.(1989), Économétrie des variables qualitatives. 2^d édition. Économica.

Greene, W.H. (1997), Econometric analysis, Third edition, Prentice Hall international editions.

Griliches, Z. (1990), "Patent statistics as economic indicators: a survey", Journal of Economic Literature, 28, pp. 1661-1707.

Iosso, T.R. (1993), "Industry Evolution with a Sequence of Technologies and Heterogeneous Ability: A Model of Creative Destruction," Journal of Economic Behavior and Organization., 21, pp109-129.

Jaffe, A. (1989), "Real Effects of Academic Research", American Economic Review, 79, pp. 957-970.

Kieffer, N.(1988), "Economic Duration Data and Hazard Functions". Journal of Economic Literature, pp. 646-679.

Le Bas, C., Picard, F., Suchecki, B. (1998), "Innovation technologique, comportement de réseaux et performances: une analyse sur données indivuduelles", Revue d'Economie Politique, 108(5), septembre- octobre 1998, pp. 625-644.

Lhuillery, S. (1996), "L'innovation dans l'industrie manufacturière française: une revue des résultats de l'enquête communautaire sur l'innovation." in Innovation, brevets et stratégies technologiques, OCDE.

Malerba, F., Orsenigo, L. (1995), "Schumpeterian Patterns of Innovation", Cambridge Journal of Economics, 19, pp. 47-65.

Mansfield, E. (1986), "Patents and innovation: an empirical study", Management science, (February), pp.173-181.

Metcalfe, J.S.(1993), "Some Lamarkian Themes in the Theory of Growth and Economic Selection", Revue internationale de systémique, vol. 7, n°5, pp487-504.

Metcalfe, J.S., Gibbons, M. (1986), "Technological Variety and the Process of Competition.", Économie Appliquée, n°3, pp 493-520.

Nelson, R., Winter, S. (1982), An Evolutionary Theory of Economic Change, Harvard University Press.

Pakes, A. (1985), "On Patents, R&D and the Stock Market Rate of Return," Journal of Political Economy, 93, pp. 390-409.

Patel, P. (1995), "Localised Production of Technology for Global Markets," Cambridge Journal of Economics, 19(1), pp. 141-154.

Patel, P., Pavitt, K. (1994), "The Continuing, Widespread (and Neglected) Importance of Improvements in Mechanical Technologies," Research Policy, 23, pp. 533-546.

Patel, P., Pavitt, K. (1995), "Patterns of technological Activity: Their Measurement and Interpretation", in P. Stoneman. (ed.), Handbook of the Economics of Innovation and Technical Change, Basil Blackwell, Oxford.

Pavitt, K. (ed.), Technical Innovation and British Economic Performance, Macmillan, 1980.

Pavitt, K., Soete, L. (1980), "Innovative Activities and Export Shares", in K. Pavitt (ed.), Technical Innovation and British Economic Performance, Macmillan, 1980.

Saviotti, P. (1996), "Technology evolution, variety and the economy," Cheltenham and Bookfield,MA. Edward Elgar.

Scherer, F. M. (1986), Innovation and Growth: Schumpeterian Perspectives, Cambridge: MIT Press.

Schmookler, J. (1966), Invention and Economic Growth, Harvard University Press.

Soete, L (1981), "A general test of the technological gap trade theory", Weltwirtschaftliches Archiv, 117(4), pp. 638-60.

Teece, J., Pisano, G. (1994), "The dynamic Capabilities of Firms: An Introduction," Industrial and Corporate Change, 3, pp. 537-556.

Tirole, J. (1988), The Theory of Industrial Organization. MIT Press, Cambridge, MA.

Winter, S.G. (1984), "Schumpeterian Competition in Alternative Technological Regimes," Journal of Economic Behaviour and Organization, 5, pp. 287-320.
ur and Organization, 5, pp. 287-320.

Chapter 3

FACTORS OF ENTRY AND PERSISTENCE IN INNOVATION:
A COMPETENCE BASED APPROACH

Alexandre Cabagnols, *Clermont-Ferrand University*
and University of Lyon 2

1. INTRODUCTION

Results about persistence reported in this book lead to two main conclusions: firstly most of the patenting firms only patent for one year, afterwards they never or very sporadically patent; secondly, a few firms succeed in continuing to patent for long periods of time. Unfortunately patent data are not sufficient to understand in detail why some firms persist in innovation while others are absolutely non-innovative or only sporadic innovators. The purpose of this paper is to explore this question with qualitative data collected in two French surveys devoted to study innovation and the competences for innovation. More precisely we will address three questions: 1- What are the competences that explain entry into innovation? 2- What are the competences that explain persistence in innovation? 3- What are the differences between competences that promote entry and those that promote persistence? This chapter is structured into three parts. Firstly we briefly discuss the relevance of this questioning and we describe the organization of our empirical analysis. Secondly we present the French surveys on which the analysis is based and we report basic descriptive results. Thirdly we comment on the results obtained from different binary logit estimations in which competences are entered as explanatory variables of the probability of entry and persistence in innovation. In conclusion we discuss the scope and implications of the results.

2. COMPETENCES, ENTRY AND PERSISTENCE

2.1 CONTEMPORANEOUS VERSUS DELAYED IMPACT OF THE COMPETENCES ON THE INNOVATIVE PROCESS

In the evolutionary tradition the concept of competence is grounded in the notion of learning and technological trajectory. If competences are a firm's specific assets it is because learning is local and cumulative (Nelson and Winter (1982); Nelson (1994), Malerba (1992)). Once we control for industry specific technological regimes, firms' competences are considered to be one the most important sources of intra-sectoral inter-firm heterogeneity and, consequently, one of the main explanatory variables of the firms' ability to innovate, grow and earn money (Audretsch (1995), Waring (1996), Foray, Mairesse (1999), Cefis (1999),). At the moment most of the empirical studies report strong evidence that the level and structure of the competences owned by firms have a significant impact on their ability to innovate at a given period of time (Leiponen (1997), (2000)). However, as it is noticed by several empirical analyses based on patent data, in most of the industrial sectors, new innovators (new patentees) have a very low level of persistence in innovation whereas a few firms remain persistently innovative and account for the major part of patent applications (Malerba et al. (1997), Cefis, Orsenigo (2001)). From the point of view of the evolutionary competence theory this conclusion is somewhat puzzling. Since competences are usually considered to be a firm's slowly-evolving specific assets which are subject to learning-by-using, firms that own and use these assets, and moreover innovate should also benefit from a persistent advantage over their competitors and remain innovative. The answer to this question may lay in the existence of two sets of competences that both have a positive impact on the contemporaneous probability of innovation but not the same delayed impact on the probabilities of entry and persistence in innovation:

The first set of competences would determine the ability of a firm to seize environmental opportunities of innovation whose emergence would remain largely random or at least unsystematic from its point of view. The instantaneous impact of these competences on the probability of innovation would be positive (i.e. firms which use these competences would have higher probabilities of innovation than those that do not use them). Likewise non-innovative firms which use these competences would have higher probabilities of entry into innovation in the future than the firms which were also previously non-innovative but did not employ them. On the contrary, once a firm is innovative, the use of these competences may not be useful and possibly detrimental to its capacity of persistence in innovation. Consequently these competences would result in sporadic innovative behaviors.

The second set of competences would explain the ability of a firm to generate and to take advantage by itself of a stable stream of opportunities. The main property of these "stabilizing" competences would be their positive impact on the ability of the firms to enter into a process of institutionalization of the innovative process. The randomness of the innovation would be significantly lower for these firms than for firms with opportunistic competences. Innovative firms that intensively use these competences would have a higher probability of persistence in innovation than the firms which use "opportunistic" competences.

Considering entry and persistence, each competence can have three different impacts: a positive impact, no impact, a negative impact. Table 1 below summarizes these 9 possibilities.

Table 1: Classification of the competences depending on their impact on the probability of entry and persistence in innovation

		Impact on the probability of persistence		
		Negative	**Neutral**	**Positive**
Impact on the probability of entry	Negative	**Anti innovative** competences	**Entry impeding** competences	**Conflicting** persistence enhancing competences
	Neutral	**Persistence impeding** competences	**Neutral** competences	Competences **specific to persistence**
	Positive	**Conflicting** entry promoting competences	Competences **specific to entry**	**Unambiguously favorable** competences

Two extreme scenarios are possible. The first one would be a total concentration of the competences along the first diagonal. This result would indicate that the processes of entry into innovation and persistence in innovation are based on similar competences. Consequently, the low level of persistence in innovation of new innovators could not be explained by a difference in the nature of the critical competences required for entry and persistence in innovation (however it may be a problem of level of competence). The second extreme case would be an empty first diagonal with many conflicting competences or impeding competences (either entry impeding or persistence impeding). This result would be the sign of a qualitative opposition between entry promoting and persistence enhancing competences. If we consider that the switch from one set of competence to another is costly and time-consuming, a new tentative interpretation of the low level of persistence in innovation of new innovators could be suggested: "the low level of persistence in innovation of new innovators may result from the difficulties they face to switch from entry promoting competences to persistence enhancing competences". To confirm this hypothesis an additional examination of the population of new innovators is however necessary (it is not carried out in this study). Its purpose would be to identify clearly the reason why new innovators have not (or have) developed subsequent innovations: Is it an inability to develop new competences? Is it the failure of the previous innovations? Is it the result of strategic considerations?

2.2 PROCEDURE OF STATISTICAL INVESTIGATION

In a first section we present the usual way of measuring the impact of the competences on the probability of innovation. This methodology is not suited to study the impact of the competences on the evolution of the innovative behavior through time. Consequently, the second section presents the specific methodology that we will apply to measure and compare the impact of the competences held by firms on their probabilities of entry and persistence in innovation.

STANDARD MODELS OF COMPETENCE AND INNOVATION

Usual models of innovation investigate the contemporaneous impact of the competences on the probability of innovation. They study the relationship that exists during a given period of time t between the set of competences C_t held by firms and their technological activity I_t. Additional explanatory variables are often included in order to control for environmental factors (X_t). This kind of specification will later be referred to as Model 1 and the resulting estimated coefficients will be considered as a measure of the "static" impact of the competences on the probability of innovation.

In the simplest case when I_t takes two levels the analysis of the innovative process is made with a logit[1] model of the following form:

$$\frac{P(I_t = 1)}{P(I_t = 0)} = e^{\beta.C_t + \alpha.X_t} \text{ with the constraint}: P(I_t = 0) = 1 - P(I_t = 1)$$

$$\Leftrightarrow P(I_t = 1) = f(\beta.C_t + \alpha.X_t) = \frac{e^{\beta.C_t + \alpha.X_t}}{1 + e^{\beta.C_t + \alpha.X_t}}$$

(Model 1)

Where:
- I_t: {0; 1} where 0 stands for non innovation and 1 stands for innovation
- $P(I_t=1)$ is the probability of innovation in time t. $P(I_t=0)$ is the probability of non-innovation in time t.
- C_t is the set of competences of which we try to evaluate the impact on the innovation.
- Xt is a set of control variables (we will use control variables to take account of sectoral technological opportunities and firm size)
- αs and βs are unknown parameters that measure the impact of the explanatory variables (competences and control variables) on the probability of occurrence of an innovative behavior. The estimated values of α and β will be noted $\hat{\alpha}$ and $\hat{\beta}$ respectively.
- f() is a link function between the explanatory variables and the probability of innovation. Later we will use a binary logit function. The interest of that function is its ease of generalization and estimation in the unordered multinomial case (i.e. when I_t has more than two levels).

DELAYED IMPACT OF THE COMPETENCES

General methodology

As indicated earlier we will consider that a specific competence has a static impact on the innovation if its presence in time t affects the probability of innovation of the firm in t. In contrast a competence will be considered as having a delayed or "dynamic" effect if its presence modifies the ability of firms to be innovative in the future. The meaning of this future innovative activity is different depending on the initial status of the firm: If a firm was previously innovative, the occurrence of an innovation in the future signals a persistent innovative behavior; if a firm was not previously innovative the occurrence of an innovation in the future is an indication of entry into innovation. Given data constraints we will only consider two successive periods of time for which we measure firms' innovative behavior. As indicated in Table 2 below, in a binary context four transitions are possible between two successive periods of observation.

Table 2: Transition matrix between innovative behaviors

		Innovative activity in time t+1	
		No innovation (0)	Innovation (1)
Innovative activity in time t	No innovation (0)	Persistent non-innovator $P_{0,0}$	New innovator (entry) $P_{0,1}$
	Innovation (1)	Sporadic innovator (exit) $P_{1,0}$	Persistence $P_{1,1}$

Where $P_{i,j}$'s in the table above stand for the probability of transition from state i in t to state j in t+1. Row sum of the transition probabilities is equal to 1 and each row represents a specific binomial distribution.

This analysis can be enriched if we make a distinction between product innovation only, process innovation only and product and process innovations simultaneously (thereafter product&process). This distinction can be made in two ways:

– Distinction between types of innovative behaviors in time t and time t+1 which leads to the study of a square 4 by 4 transition matrix. In this case technological trajectories have to be modeled with a multinomial distribution of probabilities. This approach is similar to that of the 2 by 2 case and corresponds to the analysis of the transition matrix of a first order Markov chain if only two periods are taken into account.

– Distinction between types of innovative behaviors in time t and, in t+1, a simple binary distinction between innovative and non-innovative firms. As indicated in the table below this latter case corresponds to the analysis of four binomial distributions (one for each category of firm in time t).

The interest of the distinction between product, process and product & process innovators is that the competences that influence entry into and exit from innovation are likely to be quite different and maybe contradictory so that the absence of distinction may lead to biased conclusions concerning the impact of the investigated competences. To simplify the analysis and avoid the sparseness of the observations across a large number of possible transitions we will consider only the 4 by 2 transition matrix that is represented in Table 3

Table 3: Transition matrix between innovative behaviors with distinction between profiles of product, process and product & process innovators

		Innovative activity in time t+1	
		No innovation (0)	Innovation (1)
Innovative activity in time t	No innovation (0)	Persistent non-innovator $P_{0,0}$	New innovator *(entry)* $P_{0,1}$
	Product Innovation (1)	exit $P_{1,0}$	Persistence $P_{1,1}$
	Process Innovation (2)	exit $P_{2,0}$	Persistence $P_{2,1}$
	Product &process Innovation (3)	exit $P_{2,0}$	Persistence $P_{3,1}$

Model specification

In relation to our initial questioning the purpose of this statistical work is to compare the impact of the competences on the probabilities of entry and persistence in innovation. This question can be studied by the estimation of a simple binary logit model whose explanatory variables are nested by level of the initial type of innovative activity. Consequently for each explanatory variable we will estimate one parameter per initial type of technological activity. For example if we make a distinction only between innovative and non-innovative firms in t this method results in the estimation of two coefficients for each explanatory variable, one for firms that were initially innovative and a second one for those that were initially not innovative. These different estimated parameters can be statistically compared to each other. If they are not significantly different we will conclude that the static and dynamic impacts of the variables under consideration are likely to be similar. On the contrary, if these estimated parameters are significantly different we will conclude that the static and dynamic impacts of the variable are certainly different.

Models that we will estimate have the following form:

$$P(I_{t+1} = innovation) = f(\beta.C_t(I_t) + \alpha.X_{t+1}(I_t)) \quad \textbf{(Model 2)}$$

Where:

- $X_{t+1}(I_t)$ indicates that for each control variable one parameter has been estimated for each level of the initial type of innovative behavior I_t

- $C_t(I_t)$ indicates that for each competence one parameter has been estimated for each level of the initial type of innovative behavior I_t

Two variants will be considered. A first one (model 2-a) in which we makes a distinction between firms that were previously innovative and non-innovative in t (I_t:{innovation, non-innovation}). I_t will then be possible to perform a comparison between the impact of each competence on the probability of entry into innovation of firms that were not previously innovative and on the probability of persistence in innovation of the firms that were already innovative. The second model (Model 2-b) takes account of the past type of technological activity of the firms in t by introducing a distinction between firms that were previously product, process and product & process innovators (I_t:{non-innovation, product innovation only, process innovation only, product & process innovation}).

We notice that control variables will be measured during the contemporaneous period of analysis that is t+1 (X_{t+1}) whereas we will measure the delayed impact of the competences held by firm in t on their probability of persistence (or entry) in t+1. This delay is a means of evaluating the constraints imposed and/or the opportunities offered by the previous competences of the firms on their ability to innovate now. This is a way of identifying the competences which durably constrain the innovative activity. These competences are certainly important sources of long-lasting inter-firm heterogeneity.

3. DATA PRESENTATION

3.1 SURVEY DESCRIPTION

We use a dataset that covers the period [1994-2000]. We draw on one French community innovation survey (CIS) plus one French survey specifically targeted to identify the competences used by firms to innovate.

CIS SURVEY

The third French version of the European CIS survey (CIS3) was performed in 2001. Firms were questioned in relation to their technological activity over the former 3 years (1998, 1999, 2000). This survey was only carried out in firms with more than 20 employees. Among others, it was carried out on a representative sample of 5,500 manufacturing firms. The response rate of manufacturing firms was 86% which represents 89% of the turnover of this sector.

THE "COMPETENCE" SURVEY

The "Competence" survey is a French survey specifically targeted to investigate competences. It was performed in 1997 and covers the period [1994-1996]. It was sent to a representative sample of 5.000

manufacturing firms with more than 20 employees. The response rate was 83%.

In these two surveys, weights are available in order to correct the sampling procedure. In the case of the CIS3 survey, weights have also been corrected in order to avoid any non-response bias.

3.2 MEASUREMENT ISSUES

DETECTION OF THE INNOVATIVE FIRMS

In these two surveys specific questions make it possible to identify the type of innovative behavior developed by firms: non innovation, product innovation only, process innovation only and product and process innovation simultaneously. Questions used for this purpose in CIS3 and Competence are nearly the same. The typical question is:

"During the last three years, 199_{t-1}, 199_{t-2}, 199_{t-3}, did your enterprise introduce ..."

- Q1: any product that was innovative from a technological point of view?   (0/1)
- Q2: any process that was innovative from a technological point of view?   (0/1)

Firms can then be classified according to the following rule: If Q1=0 and Q2=0 then "Non innovator"; if Q1=1 and Q2=0 then "Product innovator"; if Q1=0 and Q2=1 then "Process innovator"; if Q1=1 and Q2=1 then "Product & process innovator".

In the case of CIS3, the question is slightly different from this "standard" question since it does not mention the "technological" aspect of the innovation. An additional question is asked that distinguishes between "technological innovation", "aesthetic and packaging innovations", "service innovations", "simple variation without technological content". Later in the paper we consider only firms that declare that their new products and/or new processes were innovative from a technological point of view. With this slight modification, firms' declarations in the CIS3 survey can be compared to those made in the "Competence" Survey.

SURVEY MERGING, QUALITY OF THE FIRMS' FOLLOW-UP AND ENTRY/PERSISTENCE INDICATORS

These two surveys study the same parent population of French manufacturing firms with more than 20 employees. Consequently some firms have been (randomly) included in these two successive surveys. Since firms are uniquely identified with a national id number, the merging of the observations between surveys does not raise technical problems. Table 4 reports the distribution of firms by type of innovative activity in each sample. We observe that the distributions obtained with the Competence survey and the CIS3 survey are significantly different. In particular the number of non-innovative firms is larger in the CIS3 survey than in the Competence survey. This difference can be explained by two factors. Firstly, as indicated earlier, questions used to identify the type of innovative behavior are not strictly identical. Secondly, changes in the economic conjuncture may explain this difference. More importantly, we can notice that the distributions obtained from the subpopulation of firms that are polled in both surveys are not significantly different from those that are obtained in the parent surveys. It indicates that our sub sample does not produce a biased representation of the innovative behavior of firms.

Table 4: Distribution of the firms by type of innovative behavior in different surveys

Survey	Period	Number of observations	Raw number of innovative firms	Distribution by type of technological activity (corrected percentages by survey weights)			
				% non-innovative	% product only	% process only	% product & process
COMPETENCE	[1994-1996]	3846	2249	50.68	14.7	9.99	24.6
COMPETENCE ∩ CIS3	[1992-1994]	1011	752	51.62	14.38	8.31	25.69
CIS3	[1990-1992]	5100	2678	59.78	16.92	6.81	16.49
COMPETENCE ∩ CIS3	[1998-2000]	1011	673	61.85	17.7	5.02	15.44

Since we are interested in the study of the evolution of technological behaviors through time we will centre our research on the analysis of the probabilities of transition between types of technological behaviors over the period [1994-1996] and the period [1998-2000]. Table 5 below reports raw frequencies and weighted transition probabilities between technological behaviors.

Table 5 : Transition matrix between types of technological behaviors in [1994-1996] and [1998-2000].

First row: raw frequencies Second row: weighted transition probabilities		Type of technological behavior over the period [1998-2000] (CIS3 survey)				
		Non-innovative	Product only	Process only	Product & process	Total
Type of technological behavior over the period [1994-1996] (Competence survey)	Non- innovative	151 71.5%	46 16%	21 5.7%	41 6.8%	259 100%
	Product only	42 45%	57 29.6%	5 2.9%	44 22.6%	148 100%
	Process only	48 78.5%	12 6.1%	5 5.4%	16 9.9%	81 100%
	Product& process	97 46.4%	131 18.3%	23 4.7%	272 30.6%	523 100%
	Total	338	246	54	373	1011

As indicated earlier, in order to avoid data sparseness, we will not make a distinction between product, process and product & process innovations over the period 1998-2000. We will limit our analysis to a binary distinction between innovative and non-innovative firms (see Table 6).

Table 6: Distribution of the firms by type of transition (entry/exit/persistence)

First row: raw frequencies Second row: weighted transition probabilities		Type of technological behavior over the period [1998-2000] (CIS3 survey)		
		Non-innovative	**Innovative**	
Type of technological behavior over the period [1994-1996] (Competence survey)	**Non- innovative (0)**	Persistent non-innovation 151 72%	**Entry** 108 29%	
	Product only (1)	Exit 42 45%	**Persistence** 106 55%	
	Process only (2)	Exit 48 79%	**Persistence** 33 22%	
	Product& process (3)	Exit 97 46%	**Persistence** 426 54%	
	Total	338	673	1011

Persistence (PST) in innovation measures the probability for a firm initially engaged on a specific innovative trajectory i in t to remain innovative in t+1 whatever its final type of innovative activity (product, process, product & process). The higher the level of persistence in innovation of a specific population of firms the higher the cumulativeness (innovativeness) of its process of technological accumulation. Obviously, persistence is only defined for firms that are initially innovative during the period 1994-1996 (i.e. for i:{1,2,3}). Persistence in innovation and exit are linked.

$$PST_i = p(I_{t+1} \neq 0 /_{I_t = i}) = 1 - exit_i = 1 - p(I_t = 0 /_{I_{t-1} = i})$$

Entry is the probability for a firm that is initially non-innovative during the period [1994-1996] (I_t=0) to adopt an innovative behavior during the second period [1998-2000].

Interestingly we observe that the two tables above report high probability of exit from innovation. The exit rate is 45% for product innovators, 78% for process innovators and 46% for product & process innovators. This result is not specific to the pairing of these two French surveys. As indicated in Appendix 1 the pairing of other French innovation surveys made over the period 1992-2000 reveals similar tendencies[2] : the highest persistence rate is always obtained for product & process innovators while the lowest one is always observed for process innovators. With the exception of the pairing between the Competence and the CIS3 surveys, the persistence rate of non-innovative firms in non-innovation is also always superior to the exit rate of innovative firms. It means that on average innovative firms persist more easily in innovation than non-innovative firms enter into innovation (see Appendix 1).

THE MEASUREMENT OF COMPETENCES

Theoretical background

According to Malerba and Orsenigo [1993] p.51 competences define *"what a firm can do, shape the company's organizational structure and constrain the available menu of possible choices"*. According to this definition the measurement of competences can be indirectly made through the observation of what firms actually do since what is done reveals the existence of the capability of doing it. The "Competence" survey has been designed from this angle by J-P François (François et al. [1999]): firms are questioned about what they actually do or do not do in a selected set of areas that are of considered importance for innovation. No attempt is made to measure the quality or efficiency of this effort. The questionnaire reports only the existence (or not) of an activity in these specific areas. This procedure is interesting because it is clear and simple to implement. It has however several limitations. Firstly, items on which firms are questioned are limited to what was relevant from the point of view of the designer of the survey. Secondly firms can potentially do more than and perform differently to what they actually do. Consequently observed behaviors in this survey only reveal a subset of the whole set of competences actually held by firms. Thirdly as already indicated we do not have information concerning the intrinsic value and success of this behavior.

For the purpose of our work we have classified and accumulated elementary competences in a specific manner (different from what was suggested by François et al. [1999]). We distinguish three major sets of competences:

Competences of external interface whose structure and intensity in the firm are revealed by the existence and level of the interfaces towards:

Users (USR).

Suppliers (SUPP)

Competitors (COMP)

Global technological environment via activities of "technological watch" (WATCH)

Absorptive capabilities whose nature and strength are revealed by the existence and level of the engagement of the firms in the following areas:

R&D (RD)

Cooperation with other firms or institutions for the innovation (COOP)

Recruitment of highly qualified workers for the innovation (SCI)

Continuous training of the employees (TRAIN)

Stimulation of knowledge sharing (HOM)

Stimulation of the individual initiative and creativity (HET)

Competences of management of a global strategy of innovation whose specificity and intensity are revealed by the existence and level of the interest that firms devote to promote specific policies of:

Technological exploration (i.e. systematic encouragement of the novelty, creativity and explorative activities) (EXPLOR)

Technological knowledge accumulation (via the exploitation of past technological experiences) (ACCU)

"Rational" technological evaluation (via a systematic cost/advantage analysis of any technological project) (EVAL)

Intellectual protection (*via* secret and or legal intellectual property rights) (IP).

Appendix 2 describes in detail the construction of these variables. They are all supposed to have a positive impact on the static probability of innovation. Their impact on the probability of entry and persistence in innovation is however largely unknown. We may suppose that competences resulting in the establishment of a permanent in-house ability to innovate should have a positive impact on the probability of persistence in innovation (mainly technological watch, R&D, human resource management competences (recruitment of qualified workers, training)) whereas competences that only bring access to outside opportunities should contribute more to the probability of entry into innovation than to the probability of persistence.

Treatment of the multicolinearity

Competences used by firms are not developed independently from each other but are in a "systemic" interdependence that insures the "coherence of the firm". Consequently most of them are significantly correlated to each other which induces multicollinearity in linear estimations. To solve this problem we have performed a principal component analysis (PCA) on the scores obtained by firms on each individual competence[3]. This PCA produces 15 principal components whose interpretation is reported below. We have willingly not suppressed from the analysis the axis with little inertia since our purpose was only to reduce the multicollinearity between explanatory variables, not to limit the number of explanatory variables. Nevertheless we will be as much prudent in interpreting results as they are obtained with axes which have a little inertia. That's the reason why we will always report the results obtained with the initial variables centered by sectoral means and those obtained with a transformation of the original variables by PCA. Appendix 4 provides detailed results of this PCA Table 7 reports only a synthetic interpretation of the principal components.

Table 7: Interpretation of the 15 principal components

PRIN1	Overall high level of competence High average level of competence for innovation (mainly driven by a management style that promotes innovation, knowledge-sharing and benchmarking of competitors). Variables related to the competences of "management of a global strategy of innovation" rank particularly high on this axis. Competences of external interface (mainly towards competitors) are also well represented.
PRIN2	Market for technologies, cooperation and R&D vs. stimulation of individual creativity and knowledge sharing Capability of accessing complementary technologies through the market for technologies (MKTTEC) and cooperation (COOP) with the help of internal R&D (RD) rather than promoting the internal potential of creativity (HET) and knowledge of the employees (HOM).
PRIN3	Technological watch, suppliers and highly qualified workers Capacity to perform technological watch (WATCH), to establish external interfaces towards suppliers (SUPP) and to recruit workers with high levels of technical and scientific competences (SCI).
PRIN4	Continuous training and in-house creativity vs. inspiration by users Competence oriented towards the stimulation of the average human potential in term of skills via continuous training (TRAIN) and stimulation of individual creativity (HET) rather than towards the development of targeted competence of interface with users (USR).
PRIN5	Technological watch vs. recruitment of highly qualified workers. Ability to take advantage of external opportunities through technological watch (WATCH) vs. ability to rely on the in-house competences of highly qualified workers (SCI).
PRIN6	Interface with suppliers and the market for technologies vs. recruitment of highly qualified workers Ability to take advantage of external opportunities through interfaces with suppliers (SUPP) and the market for technologies (WATCH) vs. ability to rely on the competences of highly qualified workers (SCI).
PRIN7	Market for technologies and technological watch vs. recruiting of highly qualified workers and R&D. Ability to exploit external opportunities through technological watch and the market for technologies vs. ability to perform in house research activity based on the recruitment of highly qualified workers and R&D activities.
PRIN8	Continuous training and intellectual protection vs. cooperation and explorative management Competence of internal development of scarce competences through continuous training plus protection of the intellectual assets of the form vs. competence to establish cooperation and develop a style of management that promotes innovation and novelty.
PRIN9	Routinized technological accumulation vs. intellectual protection, creativity and benchmarking of competitors Competence of knowledge accumulation (ACCU) and knowledge evaluation (EVAL) vs. competences of intellectual protection (IP), individual creativity stimulation (HET) and interface towards competitors (COMP).
PRIN10	Cooperation and interface towards users. Competences of external interface with users (USR) and establishment of cooperation (COOP) vs. competence to use the market for technologies (MKTTEC).
PRIN11	R&D and competitors vs. intellectual protection and evaluative competences Capability of performing R&D and managing the benchmarking of competitors (COMP) vs. ability to encourage a rational cost/benefit evaluation of the technological opportunities (EVAL) and to organize the protection of the intellectual assets of the firm (IP).
PRIN12	Interface towards users vs. competitors. Capability to manage external interfaces towards users (USER) vs. towards competitors (COMP).
PRIN13	Technological accumulation and intellectual protection vs. competitors and evaluative analysis of the innovation Competence to manage a strategy of technological accumulation (ACCU) and a high level of protection of the technological assets of the firm (IP) vs. competence to manage rational cost/benefit evaluations of the technological opportunities (EVA) and to establish external interfaces towards competitors (COMP).
PRIN14	Stimulation of individual creativity vs. knowledge-sharing practices. Orientation of the competences towards the stimulation of individual initiative and creativity rather than towards knowledge sharing between workers.
PRIN15	Team level vs. business level innovative strategies. Management of the innovative process of individuals (HOM and HET) vs. that of the firm (EXPLOR).

3.3   CONTROL OF SPECIFIC FACTORS

INDUSTRIAL SPECIFICITIES

As indicated in the two graphs below (Figure 1), innovation rates and average levels of competences differ strongly across industrial sectors (detailed descriptive statistics are available in the appendix 3. These differences are explained by specific sectoral opportunity conditions that are not of direct interest for use.

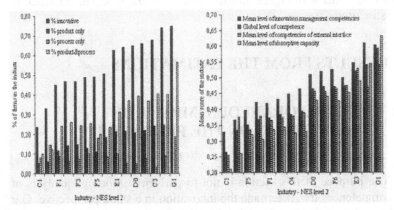

Figure 1: Illustration of the inter-industrial heterogeneity of opportunity conditions and competences
(Original data used in these graphs with full descriptors of the sectors are available in appendix 3)

In order to neutralize these sectoral specificities that have a direct impact on the probability of entry and persistence in innovation we have introduced a limited number of dummy variables at an aggregated level of the industrial classification (NES36[4]). In addition to this technique we have defined a new variable PCTI *"the percentage of non-innovating firms"* in the sector at the third and higher level of disaggregation of the NES (NES114). This variable is intended to measure the level of the technological opportunities met by firms in their specific field of activity.

To make the distinction clearer still between industrial specificities and individual characteristics, we have finally standardized firms' scores on each competence by the average score of their sector of activity defined at the third level of the NES (NES114).

FIRM SIZE

Sectoral dummy variables capture intersectoral differences of size but do not monitor possible inter-firm differences of size inside the industry. To take account of this intra-industrial heterogeneity we have constructed three new dummy variables. Sectors of reference are defined at the third level of the NES (NES114); for each of them three percentiles of turnover are defined. The first (S1) second (S2) and third (S3) dummies respectively stand for firms whose turnover belong to the first second and third percentiles of turnover in comparison to their industry. S3 that represents firms with higher turnover is used as the group of reference in the estimations.

4. RESULTS FROM THE ESTIMATIONS

4.1 CONTEMPORANEOUS IMPACT OF COMPETENCES ON THE PROBABILITY OF INNOVATION

The purpose of this section is not to perform a thorough analysis of the competences that determine the innovation in a static perspective. Our goal is only to estimate a model that will be used later as a benchmark for the interpretation of the other models. In particular we will show that a competence can have a contemporaneous or "static" impact that is positive whereas its delayed impact may be negative for specific sub-populations of firms. Here we present the results obtained after the estimation of the Model 1:

$$P\left(I_t = innovation\right) = f\left(\beta.C_t + \alpha.X_t\right)$$

Where:
  t stands for the period [1994-1996];
  X_t is the set of control variables measured in t. It is made of 3 variables:
  - LPCTI$_t$: log of the percentage of innovative firms in the sector in t
  - S1$_t$, S2$_t$, S3$_t$: dummy variables that indicate the percentile of turnover of each firm in 1994 in comparison to its industry of activity (S3 will be used as baseline).
  - Dummies for sectors defined with the NES36. The sector F5: "Manufacture of basic metals and fabricated metal products" will be used as a baseline for the estimations.
C_t is the set of competences of which we try to measure the impact on the probability of innovation. We will use two measures of the competences:
  - via a direct measure of each initial competence.

- via the coordinates of the firms on the 15 principal components obtained from the previous PCA.

In both cases the link function f() is a logit function. α's and β's are estimated with the maximum likelihood method.

Table 8 reports the results obtained after estimation of model 1 with the initial variables centered by sectoral means.

Table 8: Results from the estimation of Model 1 with the initial variables

Model 1 : Impact of the competences held in t: 1994-1996 on the probability of innovation during the period t: 1994-1996 (competences are measured with the initial variables centered by sectoral means); binary logit estimation. Endogenous variable: I_t where $I_t=1$ for innovative firms; $I_t=0$ for non-innovative firms.		Innovative activity over 1994-1996
		$P(I_t=1)/P(I_t=0)$
Estimated impact of the competences used during the period t:[1994-1996]		β
Comp. of external interface towards suppliers	SUPP	0.434
Comp. of external interface towards customers	USR	1.721*
Comp. of external interface towards competitors	COMP	0.205
Comp. of external interface technological watch	WATCH	-0.209
Comp. of cooperation with other institutions	COOP	1.206*
Comp. to manage a formal internal research activity	RD	2.455*
Capability of using the market for technologies	MKTTEC	-0.186
Comp. of stimulation of individual initiative and creativity	HET	-1.991*
Comp. of knowledge-sharing between workers	HOM	0.683
Comp. to recruit highly qualified workers for the innovation	SCI	1.13*
Comp. to organize continuous training of the employees	TRAIN	0.628*
Comp. to manage an explorative innovation strategy	EXPLOR	1.464*
Comp. to manage the accumulation of the technological experience	ACCU	1.208*
Comp. to manage a rational innovative strategy (cost/benefit analysis)	EVAL	-1.568*
Comp. to manage the protection of the intellectual assets of the firm	IP	-0.282
Control variables measured during the period [1994-1996]		α
Intercept	INTERCEPT	-2.236*
Log(Percentage of non-innovative firms in the sector)	LPCTI	-2.871*
First percentile of turnover in 1994 versus third percentile	S1	-0.202
Second percentile of turnover in 1994 versus third percentile	S2	-0.464*
+15 dummies for sectors not represented here	SECTOR	....
* **indicates significant variables at the 5% level**. Raw number of observations: 1002 (743 innovative; 259 non-innovative). -2logL intercept only=1401.543; -2logL with covariates=886.263; Degree of freedom: 32. Likelihood ratio test : 515.2*; Wald test :261*; % of concordant predictions : 88.1		

The previous table indicates that among the set of competences of external interface, the interface towards users has a positive impact on the probability of innovation. Among the absorptive capacities those that have a major role are: R&D, the recruitment of highly qualified workers (SCI) and the continuous training of the workforce (TRAIN). We notice that the competences that lead to the stimulation of individual creativity (HET) have a negative impact on the probability of innovation.

Concerning the competences of management of the innovative activity, two competences have a positive impact: the competences of stimulation of an explorative attitude towards innovation (EXPLOR) and the competences of technological accumulation via a systematic analysis of past technological experiences (ACCU). Interestingly, the competences associated to a "rational" and "evaluative" attitude towards technological change (EVAL) have a negative impact on the probability of innovation.

If we consider now the results obtained after the estimation of the model 1 with variables transformed by PCA (see Table 9) we observe a positive and significant impact of the main principal components PRIN1 that stands for the "overall level of competence of the firm". It indicates that the competences taken into account in this study have on average a positive impact on the probability of innovation. It is in line with the idea where, the higher the level of competence, the better it is for innovation. The probability of innovation is also improved by firms' ability to access complementary resources via cooperation and via the market for technologies (see the positive and significant coefficient of PRIN2).

The precise structure of the competences that firms use also matters (this point cannot be easily studied if we work with the initial competence scores). Some configurations of competences do not favor innovation. This is the case of firms that are mainly interested in developing the interface with suppliers and which intensively use the market for technologies rather than their own R&D resources (see the negative and significant coefficients associated to PRIN6 and PRIN7). The significant and positive coefficient of PRIN11 indicates that the probability of innovation increases when firms are more oriented towards R&D and the benchmarking of competitors than concerned with the stimulation of competences that promote evaluative and defensive styles of management. The orientation of the competences towards the stimulation of individual creativity rather than towards knowledge-sharing has also a negative and significant impact (see the negative coefficient of PRIN14). In the same way we observe a negative and significant impact of the competences that are oriented towards the stimulation of the innovativeness of individuals and teams rather than towards the elaboration of a business level strategy of innovativeness (PRIN15).

Table 9: Results from the estimation of Model 1 with variables transformed by PCA

Model 1: Impact of the competences held in 1994-1996 on the probability of innovation in 1994-1996 (competences transformed by principal component analysis); binary logit estimation. Endogenous variable: I_t where $I_t=1$ for innovative firms; $I_t=0$ for non-innovative firms.		Innovative activity over 1994-1996
		P(It=1) / P(It=0)
Estimated impact of the competences used during the period t: [1994-1996]		$\hat{\beta}$
High average level of competence	PRIN1	0.518*
Market for technologies, cooperation and R&D	PRIN2	0.337*
Technological watch, suppliers and highly qualified workers	PRIN3	-0.019
Continuous training and in-house creativity vs. inspiration by users	PRIN4	-0.14
Technological watch vs. highly qualified workers	PRIN5	0.074
Suppliers and the market for technologies vs. highly qualified workers	PRIN6	-0.338*
Market for techno. and techno. watch vs. recruit. of highly qualified workers and R&D	PRIN7	-0.416*
Continuous training and intellectual protection vs. cooperation and explorative management	PRIN8	-0.002
Routinized technological accumulation vs. intellectual protection and creativity	PRIN9	0.401*
Cooperation and interface towards users	PRIN10	0.337*
R&D and competitors vs. intellectual protection and evaluative competences	PRIN11	0.726*
Interface towards users vs. competitors	PRIN12	0.04
Technological accumulation and intellectual protection vs. competitors and evaluative analysis of the innovation	PRIN13	0.596*
Stimulation of individual creativity vs. knowledge-sharing practices.	PRIN14	-0.439*
Team level vs. business level innovative strategies.	PRIN15	-0.356*
Control variables measured in 1994		$\hat{\alpha}$
Intercept	Intercept	-2.236*
Log(Percentage of non-innovative firms in the sector)	Log(PCTI)	-2.871*
First percentile of turnover versus third percentile	quart94	-0.202
Second percentile of turnover versus third percentile	quart94	-0.464*
15 dummies for sectors	SECTORS	....

* indicates significant variables at the 5% level. Nb: The overall statistical characteristics of this model are strictly equivalent to that of the model estimated with the initial variables centered by sectoral means. However in this case the estimated coefficients are quite stable since the level of multicollinearity between components is null.

4.2   DELAYED IMPACT OF COMPETENCES ON THE PROBABILITIES OF ENTRY AND PERSISTENCE IN INNOVATION

The previous section has shown that the competences held in t do have an impact on the probability of innovation in t. The question we address in this section is the following: do these competences also have a delayed impact on the innovative activity? Put differently: do the competences held in t "constrain" the possibilities of innovation in the future? In particular is it possible to show that some competences have a delayed impact on the probability of entry into innovation of non-innovative firms and/or the probability of persistence in innovation of firms which were already innovative?

MODEL 2-A: DELAYED IMPACT OF THE COMPETENCES ON THE PROBABILITY OF ENTRY AND PERSISTENCE IN INNOVATION

In this model we estimate the impact of the competences held by a firm over the period 1994-1996 on its probability of innovation during the period t+1: [1998-2000] (i.e. $P(I_{t+1} = innovation)$). The model is nested within I_t where I_t stands for the type of innovative behavior developed during the period [1994-1996]. I_t takes two values: {innovation; non-innovation} so that it will be possible to estimate and compare the impact of each competence on the probability of entry into innovation for firms that were not initially innovative and of persistence in innovation for firms that were already innovative in t.

$$P(I_{t+1} = innovation) = f(\beta.C_{t+1}(I_t) + \alpha.X_t(I_t)) \qquad \text{(Model 2-a)}$$

$$P(I_{t+1} = innovation = f(\beta_0.C_i(I_t = 0) + \alpha_0.X_{t+1}(I_t = 0) + \beta_1.C_i(I_t = 1) + \alpha_1.X_{t+1}(I_t = 1))$$

Where $X_{t+1}(I_t)$ indicates that the variable X_{t+1} is nested within the variable I_t. In this case I_t takes two values: 0 for non-innovation; 1 for innovation so that we will estimate two parameters for each explanatory variable:

- One estimated parameter for firms that were not previously innovative during the period [1994-1996]. Since the endogenous variable is the probability of innovation in t+1 [1998-2000], that estimation is equivalent to the modeling the probability of entry into innovation of non-innovative firms.
- One estimated parameter for firms that were already innovative during the period [1994-1996]. That situation is symmetric to the former. In this case the estimation is equivalent to modeling the probability of persistence in innovation of firms that were already innovative.

Below are reported the results obtained after estimation of Model 2-a with the initial variables centered by sectoral means (Table 10) and with variables transformed by principal component analysis (Table 11). Both models are equivalent from a statistical point of view so that their overall properties are similar. They simply differ in their representation of the variables and in their possibilities of interpretation.

The dependent variable is the probability of innovation during the period [1998-2000]. As a whole the Model 2-a is highly significant. We have performed a test in order to evaluate the relevance of the distinction made between firms that were previously innovative and those that were not innovative during the period [1994-1996]. We have tested $H_0: \alpha_0 = \alpha_1, \beta_1 = \beta_0 \; \forall \alpha, \forall \beta$ This hypothesis is strongly rejected[5] which indicates that the impact of the explanatory variables used in that model differs certainly depending on the initial technological situation of the firm (whether it was previously innovative or not). A more precise analysis reveals that the differences between estimated coefficients are very important.

In some case the estimated coefficients have opposed signs and are both significantly different from 0 which indicate that the variables under consideration have opposite impacts on the probability of entry into innovation and on the probability of persistence in innovation.

<u>Estimations obtained with the model based on the initial variables centred by sectoral means</u>

Concerning the results reported in the model based on the initial variables centered by sectoral means (Table 10), the hypothesis H0: $\beta_1 = \beta_0 = 0$ is rejected for six competences: WATCH, EVAL, HOM, RD, TRAIN and IP. The hypothesis H0: $\beta_1 = \beta_0$ has also been tested; it is rejected for 5 out of 15 competences (WATCH, HOM, TRAIN, EVAL, IP).

Unambiguously favorable competences

The ability to perform R&D activities (RD) is the only competence with a positive impact on the probability of persistence and no significantly different impact on the probability of entry[6.] It indicates that firms' ability to perform R&D activities is quite important for persistence and certainly also for entry into innovation[7]. We classify that competence as an unambiguously favorable competence for the innovation.

Competences with a positive impact on the probability of entry and a negative impact on the probability of persistence

Two variables have a positive and significant impact on the probability of entry (WATCH and HOM) whereas in the same time they affect negatively and significantly the probability of persistence in innovation. Non-innovative firms that use technological watch and promote knowledge sharing have higher probabilities of entry into innovation whereas innovative firms which use these competences face lower probabilities of persistence in innovation.

The purpose of technological watch (WATCH) is to increase the ability of the firms to grasp technological opportunities in their environment. The consequence is a higher probability of entry into innovation for non-innovative firms that perform this kind of activity. However, the negative impact of technological watch on the probability of persistence in innovation of the firms that are already innovative indicates that this competence is certainly often related to a passive attitude of "listening" rather than to an active attitude of in-house novelty generation. The consequence may be a lower probability of persistence in innovation.

Competences of knowledge sharing (HOM) lead to a homogenization of the knowledge structure of the firm. Their positive impact on the probability of entry into innovation is certainly the consequence of an eased management of the process of technological adoption. Firstly, knowledge sharing reduces coordination costs which diminishes the costs of implementation of the innovation. Secondly shared knowledge structures surely favor the emergence of common beliefs concerning the value and necessity of the innovation which may facilitate decision making in relation to innovative projects. However, a higher homogeneity of the knowledge structure may induce a lower ability to identify and generate diversified technological opportunities. The consequence would be a low ability to sustain a continuous flow of innovative activity and a low level of persistence in innovation. That topic has been extensively studied by Cohen and Levinthal (1989).

Competences with a negative and significant impact on the probability of entry

We observe that some competences (TRAIN and IP) have a negative impact on the probability of entry whereas at the same time they tend to have a positive impact on the probability of persistence[8].

Surprisingly, the ability to perform continuous training of the workforce and the ability to manage the intellectual assets of the firm have a significant and negative impact on the probability of entry into innovation whereas they tend to have a positive (but not significant) impact on the probability of persistence in innovation[9]. For the moment we can not explain that result; further investigation based on principal components will help us understanding that phenomenon.

Two interpretations can be suggested to understand the negative impact of the competences of management of the intellectual assets on the probability of entry. The first one may be that a high score on these competences reveals bad appropriability conditions. In reaction firms increase their effort of intellectual protection. According to that explanation the negative impact of IP on the probability of entry would only reveal poor appropriability conditions (at the firm level) that discourage innovation. The second possible explanation states that a higher score on these competences reflects a defensive or risk adverse attitude towards innovation. The result would be a lower probability of innovative activity. Conversely, competences of management of the intellectual assets of the firm tend to have a positive but not significant impact on the probability of persistence in innovation which is the sign that firm level appropriability conditions matter to explain firms ability to persist in innovation.

Competences with a positive and significant impact on the probability of persistence

We observe that the competences of management of a rational innovative activity based on a systematic and rational cost/benefit evaluation of the innovative projects (EVAL) has a significant and positive impact on the probability of persistence in innovation whereas its impact on the probability of entry tends to be negative (but not significantly). The reasons of that phenomenon are clear: strong evaluative competencies may lead non-innovative firms to a more selective attitude towards innovative projects since in the absence of prior experience in innovation they undoubtedly practice high discounts rates. However this selective behavior certainly leads to the adoption of projects with higher average returns which, afterwards, surely increases the probability of persistence in innovation.

Table 10: Results from the estimation of Model 2-a with the initial variables centered by sectoral means

Model 2-a estimation performed with the initial variables centered by sectoral means. Impact of the competences held in t: [1994-1996] on the probability of innovation in t+1: [1998-2000]; binary logit estimation nested within It. It: Initial type of innovative behavior during the period [1994-1996] →		Estimation of the probability of innovation over t+1: [1998-2000] depending on the initial type of innovative activity in t:[1996-1998]		
		P(L-n=1)/P(L-a=0)		
		Non-innovation (entry)	Innovation (persistence)	Test of
Estimated impact of the competences		$\hat{\beta}_0$	$\hat{\beta}_1$	$\beta_0=\beta_1$
Comp. of external interface towards suppliers	SUPP	-0.304	0.066	ns
Comp. of external interface towards customers	USR	-0.342	0.054	ns
Comp. of external interface towards competitors	COMP	-1.094	-0.52	ns
Comp. of external interface technological watch	WATCH	2.149*	-1.483*	*
Comp. of cooperation with other institutions	COOP	0.433	-0.334	ns
Comp. to manage a formal internal research activity	RD	1.19	1.087*	ns
Capability of using the market for technologies	MKTTEC	0.973	-0.018	ns
Comp. of stimulation of individual initiative and creativity	HET	-0.852	0.053	ns
Comp. of knowledge-sharing between workers	HOM	1.262*	-1.698*	*
Comp. to recruit highly qualified workers for the innovation	SCI	1.117	-0.306	ns
Comp. to organize continuous training of the employees	TRAIN	-0.918*	0.737	*
Comp. to manage an explorative innovation strategy	EXPLOR	1.217	0.247	ns
Comp. to manage the accumulation of the technological experience	ACCU	1.537	-0.205	ns
Comp. to manage a rational innovative strategy (cost/benefit analysis)	EVAL	-1.114	2.228*	*
Comp. to manage the protection of the intellectual assets of the firm	IP	-1.643*	0.768	*
Control variables measured in 1998		$\hat{\alpha}_0$	$\hat{\alpha}_1$	
Intercept	Intercept	-1.5*	-2.235*	
Log(Percentage of non-innovative firms in the sector) (1998-2000)	Log(PCTI)	-1.179	-2.242*	
First percentile of turnover versus third percentile (1998)	S1	-0.944*	-1.3*	
Second percentile of turnover versus third percentile (1998)	S2	-0.312	-0.016	
+15 dummies for sectors not represented here (two coefficients per sector)	SECTORS	....	....	

* **indicates significant variables at the 5% level.** Raw number of observations: 1009 (670 innovative; 339 non-innovative); -2logL with the intercept only: 1346.3; -2logL with covariates: 924.4; Degree of freedom: 66. Likelihood ratio test : 439.3* ; Wald test : 245* ; % of concordant predictions : 76.8

Estimations obtained with the model based on variables transformed by principal component analysis

Overall properties of this model are similar to those of the model estimated with the initial variables centered by sectoral means but it leads to somewhat different interpretations since the explanatory variables are perfectly orthogonal and classified by decreasing inertia. Results are reported in Table 11. From a general point of view 8 principal components out of 15 have a significant impact on the probability of entry and persistence[10]: PRIN1, PRIN2, PRIN3, PRIN7, PRIN8, PRIN9, PRIN13 and PRIN14.

Table 11: Results from the estimation of Model 2-a with variables transformed by PCA

Model 2-a: **Estimation with variables transformed by PCA**: Impact of the competences held in t: [1994-1996] on the probability of innovation in t+1: [1998-2000]; binary logit estimation nested within It.	Estimation of the probability of innovation over t+1: [1998-2000] depending on the initial type of innovative activity in t: [1996-1998] $P(I_{t+1}=1)/P(I_{t+1}=0)$		
	Non-innovation (entry)	Innovation (persistence)	
I_t: **Initial type of innovative behavior during the period [1994-1996]** →			Test of
Estimated impact of the competences measured over the period [1994-1996]	$\hat{\beta}_0$	$\hat{\beta}_1$	$\beta_0=\beta_1$
High average level of competence	PRIN1	0.175*	0.108	ns
Market for technologies, cooperation and R&D.	PRIN2	0.354*	0.048	ns
Technological watch, suppliers and highly qualified workers	PRIN3	0.316*	-0.337*	*
Continuous training and in-house creativity vs. inspiration by users	PRIN4	0.251	-0.263	*
Technological watch vs. highly qualified workers.	PRIN5	0.224	-0.139	ns
Suppliers and the market for technologies vs. highly qualified workers	PRIN6	-0.099	-0.028	ns
Market for techno. and techno. watch vs. recruit. of highly qualified workers and R&D.	PRIN7	0.109	-0.435*	*
Continuous training and intellectual protection vs. cooperation and explorative management	PRIN8	-0.764*	0.494*	*
Routinized technological accumulation vs. intellectual protection and creativity	PRIN9	0.539*	0.149	ns
Cooperation and interface towards users.	PRIN10	-0.377	-0.184	ns
R&D and competitors vs. intellectual protection and evaluative competences	PRIN11	0.372	-0.213	*
Interface towards users vs. competitors.	PRIN12	0.104	0.149	ns
Technological accumulation and intellectual protection vs. competitors and evaluative analysis of the innovation	PRIN13	0.346	-0.384*	*
Stimulation of individual creativity vs. knowledge-sharing practices.	PRIN14	-0.417	0.475*	*
Team level vs. business level innovative strategies.	PRIN15	0.006	-0.371	ns
Control variables measured in 1998		$\hat{\alpha}_0$	$\hat{\alpha}_1$	
Intercept	Intercept	-1.5*	-2.235*	
Log(Percentage of non-innovative firms in the sector) (1998-2000)	Log(PCTNI)	-1.179	-2.242*	
First percentile of turnover versus third percentile (1998)	S1 vs. S3	-0.944*	-1.3*	
Second percentile of turnover versus third percentile (1998)	S2 vs. S3	-0.312 .	-0.016	
+15 dummies for sectors not represented here (two coefficients per sector)	SECTORS	...	...	

* indicates significant variables at the 5% level. Nb.: The overall statistical characteristics of this model are strictly equivalent to that of the model estimated with principal components.

Entry promoting and persistence enhancing competences

None of the principal component entered into the model has at the same time a positive and significant impact on the probability of entry and a positive and significant impact on the probability of persistence. PRIN1 which stands for the "average level of competence" of the firms and account for the major part of the inertia of the PCA (43% of the variance) may be considered as the only exception since its impact on the probability of entry is significantly positive and its impact on the probability of persistence is positive and tends to be significant[11].

The reverse situation (negative and significant impact on the probability of entry and persistence) is not observed. These two phenomena indicate that entry and persistence in innovation are not driven by similar factors.

Competences that only affect the probability of entry

Two components have a significant impact on the probability of entry but no significant impact on the probability of persistence. The first and most important component is PRIN2 (it explains 9.4% of the variance of the original dataset). It stands for firms' ability of using the market for technologies and of establishing cooperation with other institutions. This variable has a positive and significant impact on the probability of entry but no significant impact on the probability of persistence. It means that markets for technologies and cooperation may help firms enter into innovation but they are certainly not enough to stimulate a persistent innovative activity.

The second component with a positive and significant impact on the probability of entry but no significant impact on the probability of persistence is PRIN9 which stands for the ability of the firm to manage routinized process of technological accumulation rather than to manage intellectual protection and creativity. This result indicates that the exact orientation of the management of the innovation in the firm matters. A "routinized" management of the innovation (i.e. based on the ability of the firm to organize the feed-backs from previous technological experiences and to perform rational evaluations of its innovative projects) is more favorable to the entry into innovation than a management of the innovation based on a more competitive and strategic approach (i.e. focused on protecting the intellectual assets of the firm, on analyzing competitors and on stimulating individual creative behaviors).

Competencies that only affect the probability of persistence

The main component with a specific impact on the probability of persistence is PRIN7. This axis opposes on the one hand the ability of the firm to exploit the technological knowledge available in its global environment (i.e. to practice technological watch and to use the markets for technologies) and on the other hand the ability of the firm to perform in-house researches via the recruitment of highly qualified workers and R&D activities. The estimated impact of that variable on the probability of entry is not at all significant. However, its impact on the probability of persistence is significant and negative. It indicates that the probability of persistence in innovation is larger for firms that favor in-house research over technological watch and over the use of the market for technologies. The development of formal in-house R&D competences is thus an important factor of persistence in innovation that can not be balanced by technological watch and the market for technologies.

PRIN13 is the second component with a significant impact on the probability of persistence in innovation but no significant impact on the probability of entry. This component represents the ability of the firm to apply a cumulative and defensive style of management of the innovation (i.e. based on the capacity of the firm to organize the feed-backs from its past technological experiences and to protect its intellectual assets) rather than to carry out a rational and competitive management of the innovation (i.e. based on the ability to perform a rational cost/benefit analysis of its innovative projects and on the ability to monitor its competitors). The negative sign of the estimated coefficient associated to PRIN13 on the probability of persistence indicates that the higher are the efforts devoted to exploit the cumulativeness of the technological change and to protect the innovation from imitation in comparison to the efforts devoted to carry out a rational evaluation of the innovative projects and to monitor competitors, the lower is the probability of persistence. We suggest the following explanation: When firms both have the ability to perform a rational evaluation of the intrinsic value of their innovative projects and the capacity to carry-out a competitive assessment they certainly select projects which, on average, fit better to the market than the projects which are selected by the firms which are only concerned with their past technological experiences and with the protection of their intellectual assets. The consequence may be a low profitability of the innovation for firms that rank high on PRIN13 and consequently a low probability of persistence in innovation.

Competences with opposite and significant impacts on the probability of entry and persistence

Three components have opposed impacts on the probability of entry and persistence in innovation.

- PRIN3 has a positive impact on the probability of entry and a negative impact on the probability of persistence.
- PRIN8 and PRIN14 have a negative impact on the probability of entry but a positive impact on the probability of persistence.

PRIN3[12] stands for the ability of the firms to perform technological watch, to establish external interfaces towards suppliers and to recruit highly qualified workers.

The positive impact on the probability of entry indicates that outward looking competences of technological watch and interfaces with suppliers are excellent sources of opportunities for the innovation. However the negative and significant impact of PRIN3 on the probability of persistence reveals that these outward looking competences are detrimental to the probability of persistence in innovation. That phenomenon certainly reveals an antagonism between entry promoting competences based on the ability of the firm to establish external interfaces and persistence enhancing competencies which are surely more dependent on the competencies of internal technological accumulation and diversity generation of the firm. The interpretation of this result is similar to the interpretation of the estimated coefficient of PRIN7[13].

PRIN8[14] opposes competences of continuous training and intellectual protection to competences of cooperation and promotion of business level innovative policies. Its impact on the probability of entry is significantly negative. It reveals that entry is more favorably influenced by the establishment of cooperation and the development of a style of management oriented towards novelty than by policies of continuous training and intellectual protection. That's exactly the opposite for persistence: the probability of persistence is higher when firms are more inclined to stimulate continuous training and IP rather than to establish cooperation and to adopt business level innovative policies. Here again it looks like an opposition between competencies that increase the ability of the firms to benefit from momentary external technological opportunities via outward looking competencies (in that case the establishment of cooperation) and competencies that induce the development of specific and permanent in-house capabilities of innovation (in this case via continuous training).

Despite its low inertia, the meaning of PRIN14 is clear-cut. It opposes on one hand the competences dedicated to encourage individualistic creative activities which result in a more heterogeneous knowledge structure and, on the other hand, the competences which favor knowledge sharing and result in a higher level of homogeneity of the knowledge structure of the firms. The equivalent to this principal component may have been directly obtained from raw variables by simply subtracting the variable HOM to the variable HET (HET *minus* HOM). The introduction of HET *minus* HOM in a model produces exactly the same results as those that are reported here: the orientation of the competences of the firms towards the stimulation of individualistic innovative behaviors rather than towards the stimulation of knowledge sharing reduces significantly the probability of entry into innovation of non-innovative firms whereas it significantly increases the probability of persistence of innovative firms. We suggest the following interpretation:

- Innovative firms enjoy the experience gained through at least one previous period of innovation whereas non-innovative firms do not benefit from a recent innovative experience. Therefore, the implementation of an innovation represents a larger break for non-innovative firms than for innovative ones. That break is certainly easier to manage when the knowledge structure of the firm is homogeneous than when it is heterogeneous since it surely smoothes the process of decision making concerning the innovation and reduces coordination costs related to the introduction of the novelty. That phenomenon would explain the negative impact of PRIN14 on the probability of entry into innovation of non-innovative firms. The drawback of knowledge sharing practices is that they reduce the internal diversity of the knowledge structure which may lead to a lower level of creativity and to a decreased ability to identify emerging external opportunities. The consequence would be a lower level of persistence in innovation. Conversely, competences of stimulation of the individualistic innovative behaviors would make entry into innovation more complex since decision making is more difficult and coordination costs are higher in heterogeneous environments than in homogeneous ones. However, once firms benefit from the experience of a previous innovative activity and once they have developed "routines of technological change", decision making and coordination are easier which makes creativity and reactivity towards the environment more critical to explain persistence in innovation than knowledge sharing.

MODEL 2-B: DELAYED IMPACT OF THE COMPETENCES ON THE PROBABILITY OF PERSISTENCE IN INNOVATION DEPENDING ON THE INITIAL TYPE OF INNOVATIVE BEHAVIOUR

This last model takes account of the possibility for a same competence to have a varying delayed impact depending on the precise type of innovative behavior that firms have initially developed in t. This model is nested within I_t where I_t takes four levels: non-innovation, product innovation; process innovation; product & process innovation. Except for that point this model is similar to Model 2-a. Notice also that we only estimate two coefficients for each sector[15] since the estimation of 4 coefficients per sector would have induced difficulties of estimation due to the sparseness of the data across a large number of modalities.

$$P\left(I_{t+1} = innovation\right) = f\left(\beta.C_{t+1}(I_t) + \alpha.X_t(I_t)\right)$$

$$P\left(I_{t+1} = innovation\right) = f\left(\begin{array}{l} \beta_0.C_t(I_t = 0) + \alpha_0.X_{t+1}(I_t = 0) \\ + \beta_1.C_t(I_t = 1) + \alpha_1.X_{t+1}(I_t = 1) \\ + \beta_2.C_t(I_t = 2) + \alpha_2.X_{t+1}(I_t = 2) \\ + \beta_3.C_t(I_t = 3) + \alpha_3.X_{t+1}(I_t = 3) \end{array} \right) \text{(Model 2-b)}$$

Where X (I) indicates that the variable X is nested within the variable I. In this case I_t takes four values: 0 for non-innovation; 1 for product innovation only; 2 for process innovation only; 3 for product & process innovation so that we will estimate four parameters for each explanatory variable:
- One estimated parameter for firms that were not previously innovative during the period t: [1994-1996]. Since the endogenous variable is the probability of innovation in t+1: [1998-2000], that

estimation is equivalent to modeling the probability of entry into innovation of non-innovative firms.

- One estimated coefficient per explanatory variable for product innovators only (α_1's and β_1's), one estimated coefficient per explanatory variable for process innovators only (α_2's and β_2's), and one estimated coefficient per explanatory variable for product & process innovators (α_3's and β_3's). The estimated coefficients measure the impact of the explanatory variables on the probability of persistence in innovation of these three categories of firms.

We report below the results obtained after estimation of Model 2-b with the initial variables centered by sectoral means (Table 12) and with variables transformed by principal component analysis (Table 13). The model as a whole is highly significant. We have performed a test in order to evaluate the relevance of the distinction made between the different types of innovative behaviors during the period [1994-1996]. We have tested $H_0: \beta_1 = \beta_2 = \beta_3 \ \forall \beta$ This hypothesis is rejected[16] which indicates that the impact of the competences differs depending on the specific type of initial technological activity of the firm (whether it was product, process or product & process innovator).

Estimations obtained with the model based on the initial variables
centered by sectoral means

Table 12 has been obtained after estimation of model 2-b with the initial variables centered by sectoral means. The overall conclusions obtained with this model are close to those obtained with model 2-a. However, some specificity must be noticed:

- Firstly the explanatory variables have not a uniform impact on the probability of persistence. We have tested the hypothesis H0: $\beta_1=\beta_2$ =β_3 for each competence. With a 5% threshold it is rejected in four cases: WATCH, COOP, SCI, and ACCU. For example the competences of technological watch (WATCH) have a significant and negative impact on the probability of persistence in innovation of product innovators whereas the impact is not at all significant and significantly different in the case product & process innovators.

- Secondly the variables COOP, TRAIN and ACCU had no significant impact on the probability of persistence in model 2-a whereas they appear significant in Model 2-b. TRAIN and ACCU have a positive impact on the probability of persistence in innovation of product and product & process innovators respectively whereas COOP has a negative impact on the probability of persistence in innovation of product & process innovators.

Table 12: Results from the estimation of Model 2-b with the initial variables centered by sectoral means

Model 2-b: Estimations obtained with the initial variables centered by sectoral means. Impact of the competences held in t: [1994-1996] on the probability of innovation in t+1: [1998-2000]; binary logit estimation nested within It. It: Initial type of innovative behavior during the period [1994-1996] →		Estimation of the probability of innovation over 1998-2000 depending on the initial type of innovative activity in t $P(I_{t+1}=1)/P(I_{t+1}=0)$				
		Non-inno-vation (entry)	Product inno vation (persis-tence)	Process inno vation (persis-tence)	Prod. &proc. Inno vation (persis tence)	Test of $\beta_{1}=\beta_{2}=\beta_{3}$
Estimated impact of the competences measured over [1996-1998]		β_{0}	$\hat\beta_{1}$	$\hat\beta_{2}$	$\hat\beta_{3}$	
Comp. of external interface towards suppliers	SUPP	-0.304	-0.49	-2.462	0.259	ns
Comp. of external interface towards customers	USR	-0.342	0.743	-1.18	-0.125	ns
Comp. of external interface towards competitors	COMP	-1.094	2.844	-3.95	0.537	ns
Comp. of external interface technological watch	WATCH	2.149*	-6.546*	-2.428	-0.269	*
Comp. of cooperation with other institutions	COOP	0.433	2.587	2.623	-1.84*	*
Comp. to manage a formal internal research activity	RD	1.19	-0.05	6.595*	1.171	ns
Capability of using the market for technologies	MKTTEC	0.973	-0.426	-4.404	0.768	ns
Comp. of stimulation of individual initiative and creativity	HET	-0.852	-1.686	2.046	1.39	ns
Comp. of knowledge-sharing between workers	HOM	1.262*	-1.257	3.979	-2.79*	ns
Comp. to recruit highly qualified workers for the innovation	SCI	1.117	1.672	2.898	-2.183*	*
Comp. to organize continuous training of the employees	TRAIN	-0.918*	2.705*	-1.296	0.247	ns
Comp. to manage an explorative innovation strategy	EXPLOR	1.217	0.176	3.202	0.217	ns
Comp. to manage the accumulation of the current and past technological choices and activities via retrospective analysis	ACCU	1.537	-4.014	-2.574	3.21*	*
Comp. to manage a rational innovative strategy (cost/benefit analysis)	EVAL	-1.114	-0.164	-1.95	2.222*	ns
Comp. to manage the protection of the intellectual assets of the firm	IP	-1.643*	0.874	0.238	1.082	ns
Control variables measured in 1998		α_{0}	α_{1}	α_{2}	α_{3}	
Intercept	Intercept	-1.5*	-4.122*	-4.221*	-3.779*	
Log(Percentage of non-innovative firms in the sector) (1998)	LPCTI	-1.179	-6.11*	-1.956	-3.097*	
First percentile of turnover versus third percentile (1998)	S1	-0.944*	-1.399	-2.158	-3.275*	
Second percentile of turnover versus third percentile (1998)	S2	-0.312	0.081	-0.069	0.45	
+15 dummies for sectors not represented here [1]	SECTORS	...		...		

* indicates significant variables at the 5% level. Raw number of observations: 1009 (670 innovative; 339 non-innovative). -2logL intercept only=1346.3; -2logL with covariates=794.3; Degree of freedom: 104; Likelihood ratio test: 609.37*; Wald test: 250.2*; % of concordant predictions: 79.2.

[1] We have only estimated two coefficients per sector: one for firms that were not previously innovative, another for firms which were already innovative whatever their initial type of innovative activity.

Estimations obtained with the model based on variables transformed by
principal component analysis

The estimation of Model 2-b with variables transformed by PCA
produces the same kind of results as those reported by the estimation
based on the initial variables centered by sectoral means.

- Firstly the explanatory variables have not a uniform impact on the
  probability of persistence. We have tested the hypothesis H0: $\beta_1=\beta_2$
  $=\beta_3$ for each axis. With a 5% threshold it is rejected in five cases:
  PRIN5, PRIN6, PRIN8, PRIN10 and PRIN14.
- Secondly the variables PRIN1, PRIN4, PRIN5, PRIN6, PRIN8 and
  PRIN10 had no significant impact on the probability of persistence
  in model 2-a whereas now, they exhibit a significant impact on the
  probability of persistence of one population of innovators at least[17].
  PRIN1 which stands for the average level of competence of the
  firms has a positive impact on the probability of persistence in
  innovation of product & process innovators. PRIN5 has a negative
  impact on the probability of persistence in innovation of product
  innovators. PRIN6 has a negative impact on the probability of
  persistence in innovation of process innovators. The case of PRIN10
  is specific since it has opposite impacts on the probability of
  persistence of product innovators and product & process innovators:
  a stronger ability to establish cooperation and external interfaces
  towards users significantly increases the probability of persistence
  in innovation of product innovators whereas it significantly
  decreases the probability of persistence in innovation of product &
  process innovators.
- Lastly we notice that the impact of PRIN14 which stands for the
  ability of the firms to stimulate individual creativity rather than
  knowledge sharing has a very positive impact on the probability of
  persistence in innovation of product & process innovators. The
  impact of this variable on the probability of persistence of product
  & process innovators is significantly lager than it is on the
  probability of persistence in innovation of product innovators only.

Table 13: Results from the estimation of Model 2-b with variables transformed by PCA

Model 2-a: Estimation obtained with variables transformed by PCA. Impact of the competences held in t:[1994-1996] on the probability of innovation in t+1:[1998-2000]; binary logit estimation nested within I. L: Initial type of innovative behavior during the period [1994-1996] →		Probability of innovation over 1998-2000 depending on the initial type of innovative activity in t				
		P(I_{t+1}=1) / P(I_{t+1}=0)				Test of β_1 = β_2 = β_3
		Non-inno-vation (entry)	Product inno-vation (persis-tence)	Process innovatio n (persiste nce)	Prod.&pr oc. innovation (persisten ce)	
Estimated impact of the competences		$\hat{\beta}_0$	$\hat{\beta}_1$	$\hat{\beta}_2$	$\hat{\beta}_3$	
High average level of competence	PRIN1	0.175*	-0.132	0.177	0.326*	ns
Market for technologies, cooperation and R&D.	PRIN2	0.354*	0.2	-0.045	-0.196	ns
Technological watch, suppliers and highly qualified workers	PRIN3	0.316*	-0.998*	-0.298	-0.391	ns
Continuous training and in-house creativity vs. inspiration by users	PRIN4	0.251	0.214	0.648	-0.615*	ns
Technological watch vs. highly qualified workers.	PRIN5	0.224	-1.192*	-0.335	0.265	*
Suppliers and the market for technologies vs. highly qualified workers	PRIN6	-0.099	-0.567	-2.06*	0.37	*
Market for techno. and techno. watch vs. recruit. of highly qualified workers and R&D.	PRIN7	0.109	-0.471	-2.256	-0.146	ns
Continuous training and intellectual protection vs. cooperation and explorative management	PRIN8	-0.764*	1.268*	-1.365	0.228	*
Routinized technological accumulation vs. intellectual protection and creativity	PRIN9	0.539*	-0.31	-0.241	0.01	ns
Cooperation and interface towards users.	PRIN10	-0.377	1.362*	-0.171	-0.761*	*
R&D and competitors vs. intellectual protection and evaluative competences	PRIN11	0.372	0.2	0.053	0.065	ns
Interface towards users vs. competitors.	PRIN12	0.104	-0.425	0.731	-0.147	ns
Technological accumulation and intellectual protection vs. competitors and evaluative analysis of the innovation	PRIN13	0.346	-0.44	0.245	-0.215	ns
Stimulation of individual creativity vs. knowledge-sharing practices.	PRIN14	-0.417	-0.245	-1.073	1.197*	*
Team level vs. business level innovative strategies.	PRIN15	0.006	-0.843	0.017	-0.235	ns
Control variables		$\hat{\alpha}_0$	$\hat{\alpha}_1$	$\hat{\alpha}_2$	$\hat{\alpha}_3$	
Intercept	Intercept	-1.5*	-4.122*	-4.221*	-3.779*	
Log(Percentage of non-innovative firms in the sector) (1998)	LPCTI	-1.179	-6.11*	-1.956	-3.097*	
First percentile of turnover versus third percentile (1998)	S1	-0.944*	-1.399	-2.158	-3.275*	
Second percentile of turnover versus third percentile (1998)	S2	-0.312	0.081	-0.069	0.45	
Dummies for sectors not reported here	SECTOR	...	...			

* indicates significant variables at the 5% level. Nb.: The overall statistical characteristics of this model are strictly equivalent to that of the model estimated with the initial variables centered by sectoral means.

4.3   DO ENTRY PROMOTING AND PERSISTENCE ENHANCING COMPETENCES DIFFER?

The econometric results reported above are summarized in two tables below (Table 14 and Table 15). We observe that most of the competences do not have an equivalent impact on the probability of entry and persistence in innovation.

Table 14: Classification of the competences depending on their delayed impact on the probability of entry and persistence in innovation

		Impact on the probability of ...		
		Innovation in t	Entry in t+1	Persistence in t+1
Unambiguously favorable competences				
Comp. to manage a formal internal research activity	RD	+	~+	+
Competences with opposite impacts on the probabilities of entry and persistence				
Comp. of external interface technological watch	WATCH	0	+	-
Comp. of knowledge-sharing between workers	HOM	0	+	-
Comp. to organize continuous training of the employees	TRAIN	+	-	~+
Competences specific to persistence				
Comp. to manage the accumulation of the current and past technological choices and activities via retrospective analysis	ACCU	+	0	~+
Comp. to manage a rational innovative strategy (cost/benefit analysis)	EVAL	-	0	+
Competences with detrimental impacts either on the probability of entry or on the probability of persistence				
Comp. to manage the protection of the intellectual assets of the firm	IP	0	-	0
Comp. of cooperation with other institutions	COOP	+	0	~-
Comp. to recruit highly qualified workers for the innovation	SCIHRM	+	0	~-
Competences without significant impact on the probabilities of entry and persistence in innovation				
Comp. of external interface towards customers	USR	+	0	0
Comp. to manage an explorative innovation strategy	EXPLOR	+	0	0
Comp. of stimulation of individual initiative and creativity	HET	-	0	0
Comp. of external interface towards suppliers	SUPP	0	0	0
Comp. of external interface towards competitors	COMP	0	0	0
Capability of using the market for technologies	MKTTEC	0	0	0

+/- indicate the sign of the impact of the variable on the probability of innovation in t; on the probability of entry into innovation in t+1 of non-innovative firms in t; on the probability of persistence in innovation in t+1 of the firms which were already innovative in t.

~ indicates that a tendency is statistically observed for specific sub-populations of innovative firms (product, process or product & process). However, this impact is not statistically significant for the overall population of innovative firms.

Concerning the results obtained with the initial variables centered by sectoral means (Table 14), we notice that the only competence with a positive impact on the probability of entry and persistence in innovation is the ability to perform R&D activities[18]. Three competences have opposite impacts on the probabilities of entry and persistence: Technological watch (WATCH) and knowledge-sharing (HOM) increase the probability of entry whereas they decrease the probability of persistence; the reverse is true for the ability to organize continuous training (TRAIN). Two competences tend to increase the probability of persistence whereas they do not significantly effect the probability of entry (ACCU and EVAL). Finally the competence to manage the protection of the intellectual assets of the firm (IP) has a negative impact on the probability of entry and no impact on the probability of persistence in innovation. Conversely the capability of establishing cooperation (COOP) and recruiting highly qualified workers (SCI) would have a negative impact on the probability of persistence in innovation. The other variables have no significant impact on the probability of entry and persistence in innovation whereas in three cases out of six they have a significant impact on the contemporaneous probability of innovation[19].

Table 15: Classification of the competences depending on their delayed impact on the probability of entry and persistence in innovation

		Impact on the probability of ...		
		Innovation in t	Entry in t+1	Persistence
Unambiguously favorable competences				
High average level of competence	PRIN1	+	+	~+
Competences specific to entry				
Market for technologies, cooperation and R&D.	PRIN2	+	+	0
Routinized technological accumulation vs. intellectual protection and creativity	PRIN9	+	+	0
Competences with opposite impacts on the probabilities of entry and persistence				
Technological watch, suppliers and highly qualified workers	PRIN3	0	+	-
Continuous training and intellectual protection vs. cooperation and explorative management	PRIN8	0	-	+
Competences specific to persistence				
Stimulation of individual creativity vs. knowledge-sharing practices.	PRIN14	-	0	+
Competences detrimental to persistence in innovation				
Technological accumulation and intellectual protection vs. competitors and evaluative analysis of the innovation	PRIN13	+	0	-
Market for techno. and techno. watch vs. recruit. of highly qualified workers and R&D.	PRIN7	-	0	-
Suppliers and the market for technologies vs. highly qualified workers	PRIN6	-	0	~-
Continuous training and in-house creativity vs. inspiration by users	PRIN4	0	0	~-
Technological watch vs. highly qualified workers.	PRIN5	0	0	~-
Competences without significant impacts on the probabilities of entry and persistence				
Cooperation and interface towards users.	PRIN10	+	0	~+/~-
R&D and competitors vs. intellectual protection and evaluative competences	PRIN11	+	0	0
Team level vs. business level innovative strategies.	PRIN15	-	0	0
Interface towards users vs. competitors.	PRIN12	0	0	0

+/- indicate the sign of the impact of the variable on the probability of innovation in t; on the probability of entry into innovation in t+1 of non-innovative firms in t; on the probability of persistence in innovation in t+1 of the firms which were already innovative in t.

~ indicates that a tendency is statistically observed for specific sub-populations of innovative firms (product, process or product & process). However, this impact is not statistically significant for the overall population of innovative firms.

Regarding now the results obtained after transformation of the variables by principal component analysis (Table 15), only one axis, "the average level of competence" (PRIN1), has a positive impact on the probability of entry and persistence in innovation[20]. The other axes represent specific aspects of the structure of the competences used by firms. Two axes stimulate entry whereas they have no significant impact on the probability of persistence (PRIN2: the "capability of using the market for technologies of establishing cooperation and performing R&D" and PRIN9: the "capability of managing routinized technological accumulation rather than intellectual protection and creativity"). Two additional axes have opposite impacts on the probabilities of entry and persistence: PRIN3, the "ability to perform technological watch and establish interfaces with suppliers" increases the probability of entry but reduces the probability of persistence; PRIN8, the "ability to organize continuous training rather than to establish cooperation and to promote a explorative management", reduces the probability of entry but increases the probability of persistence. One axis (PRIN14: the "stimulation of individual creativity rather than the stimulation of knowledge-sharing") has only a positive impact on the probability of persistence whereas it has no significant impact on the probability of entry and a negative impact on the contemporaneous probability of innovation. Five axes have no significant impact on the probability of entry whereas they have a negative impact on the probability of persistence (PRIN4, PRIN5, PRIN6, PRIN7, and PRIN13)[21]. Three additional axes have no significant impact at all (PRIN11, PRIN15, and PRIN12)[22].

The preceding results (either obtained with the initial competences or with competences transformed by principal component analysis) lead to similar conclusions: entry and persistence in innovation are two distinct processes which draw on the mobilization of different (sometimes conflicting) sets of competences. On the one hand entry into innovation would be mainly favored by competences of external interface with suppliers and with the overall technological environment *via* activities of technological watch. The absorptive capacities favorable to entry would be mainly focused on the access to external complementary resources (via the market for technologies and cooperation) rather than on the internal development of novelty generation. Knowledge-sharing practices and to a lower extend R&D would play a major role in the diffusion and absorption of this externally generated technological knowledge. Finally, entry into innovation would benefit of an innovation policy mainly concerned with the "exploration" of new possibilities and the accumulation of the knowledge gained through experience. On the other hand, persistence in innovation would not clearly benefit of the ability of the firm to establish external interfaces (it may play a negative role in the case of technological watch and cooperation whereas benchmarking of competitors would have a positive impact). The absorptive capacities favorable to persistence in innovation would be mainly focused on the development of in-house creative skills (through internal R&D activities, continuous training and the stimulation of individual creativity rather than the promotion of knowledge sharing practices). Finally, persistence in innovation would benefit of an innovation policy mainly concerned with the protection of the intellectual assets of the firm and the establishment of rational cost/benefit decision making processes in relation to innovation.

5. CONCLUSION AND DISCUSSION

The purpose of this research was to explore one of the possible reasons why most of the new innovators (entrants) do not persist in innovation during long periods of time[23] whereas a few firms are able to persist in innovation during very long periods of time.

We show that the impact of the competences on the probability of entry and persistence in innovation is not uniform: a same competence can have different and sometimes conflicting impacts on theses two processes. More precisely, the innovative activity of entrants would be mostly influenced by the use of competences of external interface devoted to grasp environmental opportunities while the competences of internal interface would be mainly oriented towards the creation of an atmosphere of innovativeness and the reduction of coordination costs by the promotion of knowledge sharing competences. These two sets of competences are coherent: the exploitation of the environmental opportunities is certainly as easier as teams have been encouraged to innovation by an "explorative" business level atmosphere and prepared to face the coordination problems raised by change through the promotion of a shared knowledge structure. These innovative behaviors would be mainly "opportunistic" and therefore strongly related to the evolutions of the technological and commercial environment of these firms. The consequence would be a prevalence of sporadic pattern of innovation among new innovators. Concerning persistence in innovation, we report a very different and somewhat conflicting profile of competence. The probability of persistence would be positively influenced by the establishment of a permanent in-house capacity of innovativeness developed via R&D activities, continuous training and stimulation of individual creativity. The ability to protect the intellectual assets of the firm, to implement rational cost/benefit procedures of evaluation of the innovation would also play a positive role. Consequently, in addition to randomness of their "opportunistic" strategies of innovation, new innovators may also face difficulties to switch from an entry promoting profile of competence to a profile of competence better suited for persistence. That last hypothesis is however difficult to verify directly with our dataset since we have only performed a comparison between two independent groups: new entrants and previously innovative firms. An explicit test of this hypothesis involves the availability of a longitudinal dataset with at least three periods of observation for each firm: The first two periods would be employed to identify new innovators; the third period would be used to observe the post-entry evolution of the competences and innovative behavior of entrants firms.

ENDNOTES

[1] A probit model can also be used but, in our case, it produces nearly the same results as the logit model.

[2] In the case of process innovators the rate of exit is larger than the rate of entry in innovation of non-innovative firms. That result is specific to the pairing of the Competence and CIS3 surveys. It is certainly due to a problem of construction of the questionnaires.

[3] Beforehand scores were normalized at the industry level NES114.

[4] Sectors are defined with the French aggregated economic classification for the activities (NES). That nomenclature The NES is structured through three levels of desegregation. The first, second and third level respectively distinguish 16 sectors (NES16), 36 sectors (NES36) and 113 sectors (NES114) (see www.insee.fr).

[5] The Wald chi-square is 101.158 with 33 df (P(Wald chi-square with 15 df>62.02)<0.0001).

[6] H0: $\beta_0=\beta_1$ is not rejected.

[7] The estimated coefficient associated to $\beta_1.RD$ is not significantly positive at the 5% threshold but is "nearly" significant since under the null hypothesis H0: $\beta_1=0$ then p($\hat{\beta}_1 \geq 1.19$)=0.0512.

[8] From a statistical point of view that situation is less marked than the previous one (positive impact on entry and negative impact on persistence) since we do not observe any case in which β_0 is significantly lower than zero and simultaneously β_1 is significantly larger than zero.

[9] In both cases H0: $\beta_0=\beta_1$ is rejected.

[10] We have tested: H0: $\beta_0=\beta_1=0$ for each principal component.

[11] The estimated coefficient associated to $\beta_1.PRIN1$ is not significantly positive at the 5% threshold but tends to be significant at the 10% level under the null hypothesis H0: $\beta_1=0$ then p($\hat{\beta}_1 \geq 0.108$)=0,01.

[12] It accounts for 5.5% of the variance in the original dataset.

[13] We have shown that the probability of persistence in innovation was increased when the in-house competences of R&D of the firms were stimulated in comparison to their competences of technological watch and use of the market for technologies.

[14] It accounts for 3.5% of the variance of the original dataset.

[15] We estimate one coefficient for firms that were previously non-innovative in t and a second one for firms which were innovative in t whatever their initial type of innovative activity: product, process, product & process.

[16] The Wald chi-square is 50.9 with 30 df (P(Wald chi-square with 30 df>50.9)=0.01).

[17] The hypothesis H0: $\beta_1=\beta_2=\beta_3=0$ is rejected for PRIN5, PRIN6, PRIN8 and PRIN10 with a risk of false rejection of 5%. Concerning PRIN1 and PRIN4 notice that one of the estimated coefficients that measure the impact of on the probability of persistence is significant. Nevertheless, for these two variables the join hypothesis H0: $\beta_1=\beta_2=\beta_3=0$ can not be rejected at the 5% threshold.

[18] However, even in this case, the positive impact of R&D activities on the probability of entry into innovation is not very significant from a statistical point of view.

[19] The contemporaneous probability of innovation is positively effected by the competences to manage the external interface towards users (USR) and to implement an innovative style of management (EXPLOR) whereas it is negatively influenced by the stimulation of individual initiative and creativity.

[20] Even this case the positive impact on the probability of persistence is limited to product & process innovators.

[21] Notice that PIN13 has a positive impact on the contemporaneous probability of innovation whereas PRIN6 and PRIN7 have a negative impact on that probability.

[22] The case of PRIN10 is more complicated since its impact on the probability of persistence is positive for product innovators, non significant for process innovators and negative for product & process innovators.

[23] In the case of patentees usually not more than one or two years.

APPENDIX

APPENDIX 1: PROPORTION OF FIRMS THAT EXIT INNOVATION IN DIFFERENT FRENCH SURVEYS OVER THE PERIOD 1986-2000.

Over the period 1986-2000, several surveys have been carried out in France for the study of innovation. They were given to French industrial firms with more than 20 employees (with the exception of the Yale survey that was only carried out by firms with at least 50 employees). The merging of these surveys two by two on the basis of firms' id numbers enables us to study the evolution of the innovative behavior of French firms. The table below shows the probability of transition from one type of technological behavior in the earlier survey, to non-innovative behavior in the later survey.

We observe that on average 70% of the non-innovative firms during the first period of observation remain non innovative during the second period of observation. Among innovative firms the highest level of persistence in innovation is observed for product & process innovators with a persistence rate of 68.6% (1-0.514) while process innovators have the lowest level with an average persistence rate of 48.1%. We notice that the probabilities of transitions obtained for a same period with different surveys sometimes exhibit important differences. It indicates that such pairing between surveys has to be made with caution.

Table 16: Proportion of firms that exit innovation in 6 French surveys over the period [1986-2000]

	Percentage of non-innovative in the second period of observation								
	[1986-1990]/ [1990-1992]		[1990-1992]/ [1994-1996]				[1994-1996]/ [1998-2000]		
Techno-logical behavior during the first period of observa-tion	I90/ YALE	I90/ CIS1	CIS1/ COMP	CIS1/ CIS2	YALE/ COMP	YALE/ CIS2	COMP/ CIS3	CIS2/ CIS3	Mean
Non inno-vative	76.3%	82.7%	66.6%	75.6%	56.3%	70.1%	71.5%	75.8%	71.9%
Product inno-vator	41.4%	46.9%	28.0%	28.7%	23.2%	32.5%	45.0%	41.8%	35.9%
Process inno-vator	57.0%	62.9%	35.3%	50.7%	30.9%	41.7%	78.5%	58.5%	51.9%
Product & process innovator	33.4%	42.1%	17.8%	30.6%	22.1%	23.2%	46.4%	35.9%	31.4%

I90: stands for "Innovation survey" made in 1991. It was carried out in 1991 and covers the period [1986-1990].

YALE : stands for the French version of the Yale survey about appropriation in industry. It was performed in 1993 and covers the period [1990-1992].

CIS1 : First French Community Innovation Survey carried out in 1993. It covers the period [1990-1992].

CIS2 : Second French Community Innovation Survey carried out in 1997 and covers the period [1994-1996]

COMP : "Compétence" survey devoted to study the competences for innovation. It was done in 1997 and covers the period [1994-1996].

CIS3 : Third French Community Innovation Survey carried out in 2001. It covers the period [1998-2000].

APPENDIX 2: CONSTRUCTION OF THE VARIABLES USED TO MEASURE COMPETENCES

Competences are measured with the help of questions concerning the "behavior" of the firm in specific fields. Answers are always binary: yes/no. Question coding refers to the original survey.

XTRN : COMPETENCE OF EXTERNAL INTERFACE

SUPP: Competences of external interface towards suppliers
- C5_13_1: Do you obtain new equipment rapidly?
- C3_07_1: Do you obtain new supplies rapidly?
- C3_08_1: Do you absorb the knowledge incorporated in the new equipment and new components?
SUPP= mean of C5_13_1, C3_07_1, C3_08_1.

USR: Competences of external interface towards customers
- C2_04_1: Do you analyze the types of client and their needs?
- C2_05_1: Do you ask after-sales services or distributors for their clients' reactions?
- C2_08_1: Do you identify the needs or the behavior of pioneer consumers?
USR = mean of C2_04_1, C2_05_1, C2_08_1.

COMP: Competences of external interface towards competitors
- C2_01_1: Do you analyze competitive products?
- C2_02_1: Do you analyze patents registered by competitors?
- C2_03_1: Do you analyze competitors' publications produced by their engineers?
- C4_08_1: Do you evaluate your collective production of knowledge compared with your competitors?
- C5_01_1: Do you know your competitors' technology?
COMP = mean of C2_01_1, C2_02_1, C2_03_1, C4_08_1, C5_01_1.

WATCH: Competences of external interface towards the global technological environment
- C5_02_1: Do you know the technologies of the future? (Technology watch)?
- C5_02_2: Is this technology watch organized with specific procedures?
WATCH = mean of C5_02_1, C5_02_2

XTRN = mean of USR, SUPP, COMP, WATCH.

ABSORB: ABSORPTIVE CAPACITIES

MKTTEC: Capability of using the market for technologies
- C5_05_1: Do you sub-contract or do you obtain R&D?

- C5_08_1: Do you use inventions from a third-party (patents, licences)?
MKTTEC =mean (C5_05_1,C5_08_1);

COOP: Competences of cooperation with other institutions
- C5_06_1: Do you carry out R&D with other firms?
- C5_07_1: Do you carry out R&D with public R&D institutions?
- C5_11_1: Do you take part in joint-ventures, in strategic alliances or other forms of cooperation with innovation in mind?
COOP = mean of C5_06_1, C5_07_1, C5_11_1.

RD: Competence to manage a formal internal research activity
- C5_04_1: Do you carry out R&D?
- C5_04_2: Do you carry out R&D with specific procedures?
RD= mean of C5_04_1, C5_04_2.

HET: Competences of stimulation of individual initiative and creativity
- C4_02_1: Do you allow each person a certain degree of autonomy to innovate?
- C4_03_1: Do you value, during individual evaluation, originality and creativity?
- C4_09_1: Do you value the contribution of each person to the production of knowledge?
- C7_04_1: Are you transparent about each person's evaluation and about rewards for the best?
HET = mean C4_02_1, C4_03_1, C4_09_1, C7_04_1.

HOM: Competences of knowledge-sharing between workers
- C1_07_1: Do you encourage each employee to have a global vision of the firm?
- C3_02_1: Do you involve all departments in projects from the start?
- C3_04_1: Do you encourage team or collaborative work in order to innovate?
- C3_05_1: Do you encourage mobility between departments in order to innovate ?
- C4_07_1: Do you have a procedure for pooling knowledge?
- C7_03_1: When recruiting, do you evaluate the capacity to work in a team?
HOM = mean of C1_07_1, C3_02_1, C3_04_1, C3_05_1, C4_07_1, C7_03_1.

SCI: Competence to recruit highly qualified workers for innovation
-C5_09_1: Do you recruit employees with high scientific qualifications in order to innovate?
- C7_01_1: Do you know the current and future specialists on the market?
SCI = mean of c5_09_1, C7_01_1.

TRAIN: Competence to organize continuous training of the employees
- C7_06_1: Do you evaluate each person's training needs?
- C7_07_1: Do you give each person the necessary information to request and choose suitable training?
TRAIN = mean of C7_06_1, C7_07_1.

ABSORB = mean of MKTTEC, COOP, RD, SCI, TRAIN, HET, HOM.

MNGT: CAPABILITY TO MANAGE A GLOBAL BUSINESS INNOVATIVE STRATEGY

EXPLOR: Competence to manage an explorative innovation strategy
- C3_01_1: Do you structure the firm according to innovation projects?
- C4_01_1: Do you incite the formulating of new ideas?
- C4_04_1: Do you accept creativity which is not directly productive?
- C4_05_1: Do you reward original ideas when they are used?
- C1_02_1: Do you evaluate technologically the products that the company is likely to produce?
- C1_03_1: Do you evaluate the processes that the company is likely to adopt?
- C1_04_1: Do you evaluate the organization that the company is likely to adopt?

EXPLOR= mean of C1_02_1, C1_03_1, C1_04_1, C3_01_1, C4_01_1, C4_04_1, C4_05_1.

EVAL: Competence to manage a rational innovative strategy
- C2_07_1: Do you carry out final consumer trials?
- C3_03_1: Do you test the product or the new process in its operational context?
- C8_01_1: Do you anticipate the whole of the costs linked to an innovation?

EVAL = mean of C2_07_1, C3_03_1, C8_01_1.

ACCU: Competence to manage the accumulation of the current and past technological choices and activities
- C1_01_1: Do you monitor the quality and the efficiency of production?
- C2_06_1: Do you use the product as a source of information on the satisfaction of consumers?
- C3_06_1: Do you analyze the defects and failures of new processes?
- C8_02_1: Do you evaluate a posteriori the cost of former innovations?

ACCU = mean of C1_01_1, C2_06_1 , C3_06_1, C8_02_1.

IP: Competence to manage the protection of the intellectual assets of the firm
- C6_02_1: Do you choose to register or not register an item of industrial property according to the global profits of the company?
- C6_03_1: Do you integrate the risks of copying or imitation as soon as the products or processes are created?
- C6_04_1: Do you keep a close watch on the distribution of copies and imitations?
- C6_05_1: Do you fight copying and imitation judicially?
- C6_06_1: Do you react in a way to undermine the value of copies and imitations in the eyes of customers and distributors.
- C6_07_1: Can you identify your strategic knowledge and know-how?
- C6_09_1: Do you inform the personnel of the strategic and confidential character of this knowledge?

C6_10_1: Do you monitor communication concerning strategic knowledge?

IP = mean of C6_02_1, C6_03_1, C6_04_1, C6_07_1, C6_09_1, C6_05_1, C6_06_1, C6_10_1.

MNGT = mean of EXPLOR, EVAL, ACCU, IP.

Mean level of competence: **COMPETENCE**= mean of XTRN, ABSORB, MNGT.

APPENDIX 3: THE INDUSTRIAL SPECIFICITIES OF THE INNOVATIVE CONTEXT

Table 17: Types of innovative behaviors and mean level of competence in different industrial sectors (NES level 2) in the Competence survey over the period [1994-98]

NES level 2	Industry	Nb. Obs.	% inno-vative	% product only	% process only	% product & process	Mean level of competence
C1	Mfr of clothing articles & leather products	324	0.24	0.05	0.08	0.10	0.26
C2	Publishing, printing and reproduction of recorded media	281	0.33	0.06	0.15	0.12	0.33
F1	Mining & quarrying except energy prod. materials, manuf. of other non-metallic mineral prod.	237	0.45	0.12	0.09	0.24	0.37
C4	Mfr of domestic equipment	257	0.47	0.14	0.06	0.26	0.38
F3	Mfr of wood, wood products, pulp, paper & paper products	241	0.47	0.15	0.08	0.24	0.37
F2	Manufacture of textiles	236	0.49	0.13	0.10	0.25	0.39
F5	Mfr of basic metals & fabricated metal products	562	0.49	0.11	0.18	0.20	0.35
E2	Manufacture of metal products, machinery and equipment	591	0.51	0.18	0.09	0.23	0.38
E1	Building of ships & boats, manufacture of railway locomotives, rolling stock	79	0.63	0.21	0.10	0.31	0.48
F4	Mfr of chemicals, rubber, plastic & chemical products	394	0.64	0.22	0.05	0.37	0.47
D0	Mfr of motor vehicles	128	0.65	0.21	0.05	0.39	0.46
F6	Mfr of electric & electronic components	182	0.66	0.22	0.08	0.37	0.47
C3	Mfr of pharmaceuticals products, perfumes, soap & cleaning preparation	134	0.68	0.24	0.03	0.40	0.54
E3	Mfr of electric & electronic equipment	219	0.74	0.25	0.09	0.40	0.52
G1	Extrac. of coal, crude petro., gas & uranium; man. of coke, ref. petrol. prod., & nuclear fuel	16	0.75	0.00	0.19	0.56	0.59

Table 18: Mean levels of competence of external interface in different industrial sectors (NES level 2) in the Competence survey over the period [1994-98]

NES level 2	Industry	Nb. Obs.	XTRN	SUPP	USR	COMP	WATCH
C1	Mfr of clothing articles and leather products	324	0.26	0.19	0.41	0.30	0.12
C2	Publishing, printing and reproduction of recorded media	281	0.38	0.30	0.47	0.36	0.38
F1	Mining and quarrying except energy prod. materials, manuf. of other non-metallic mineral prod.	237	0.37	0.24	0.54	0.44	0.24
C4	Mfr of domestic equipment	257	0.38	0.31	0.55	0.43	0.22
F3	Mfr of wood, wood products, pulp, paper and paper products	241	0.38	0.29	0.54	0.44	0.25
F2	Mfr of textiles	236	0.39	0.34	0.55	0.43	0.24
F5	Mfr of basic metals and fabricated metal products	562	0.34	0.29	0.42	0.38	0.25
E2	Mfr of metal products, machinery and equipment	591	0.37	0.30	0.52	0.46	0.21
E1	Building of ships and boats, manufacture of railway locomotives, rolling stock	79	0.47	0.44	0.60	0.48	0.38
F4	Mfr of chemicals, rubber, plastic and chemical products	394	0.45	0.38	0.61	0.53	0.30
D0	Mfr of motor vehicles	128	0.46	0.41	0.64	0.48	0.30
F6	Mfr of electric and electronic components	182	0.47	0.47	0.56	0.49	0.35
C3	Mfr of pharmaceuticals products, perfumes, soap and cleaning preparation	134	0.47	0.33	0.73	0.53	0.29
E3	Mfr of electric and electronic equipment	219	0.53	0.51	0.66	0.54	0.41
G1	Extrac. of coal, crude petro., gas and uranium; man. of coke, ref. petrol. prod., and nuclear fuel	16	0.54	0.50	0.50	0.60	0.56

XTRN: mean level of external competences
SUPP: Competences of external interface towards suppliers
USR: Competences of external interface towards customers
COMP: Competences of external interface towards competitors
WATCH: Competences of external interface towards the global technological environment

Table 19: Mean levels of absorptive capacity in different industrial sectors (NES level 2) in the Competence survey over the period [1994-98]

NES level 2	Industry	ABSORB	RD	MKT	CO-OP	HET	HOM	SCI	TRAIN
C1	Manufacture of clothing articles and leather products	0.21	0.09	0.08	0.07	0.33	0.40	0.13	0.37
C2	Publishing, printing and reproduction of recorded media	0.26	0.08	0.09	0.07	0.39	0.49	0.17	0.54
F1	Mining and quarrying except energy prod. materials, manuf. of other non-metallic mineral prod.	0.34	0.25	0.24	0.17	0.39	0.54	0.15	0.60
C4	Manufacture of domestic equipment	0.33	0.25	0.21	0.13	0.43	0.54	0.19	0.53
F3	Manufacture of wood, wood products, pulp, paper and paper products	0.30	0.22	0.15	0.13	0.39	0.54	0.17	0.54
F2	Manufacture of textiles	0.33	0.26	0.17	0.19	0.45	0.54	0.18	0.51
F5	Manufacture of basic metals and fabricated metal products	0.32	0.21	0.18	0.14	0.42	0.54	0.14	0.62
E2	Manufacture of metal products, machinery and equipment	0.34	0.29	0.19	0.18	0.41	0.55	0.21	0.58
E1	Building of ships and boats, manufacture of railway locomotives, rolling stock	0.45	0.36	0.33	0.33	0.47	0.69	0.30	0.68
F4	Manufacture of chemicals, rubber, plastic and chemical products	0.44	0.43	0.29	0.31	0.47	0.63	0.26	0.70
D0	Manufacture of motor vehicles	0.43	0.40	0.32	0.25	0.46	0.64	0.24	0.69
F6	Manufacture of electric and electronic components	0.43	0.35	0.26	0.25	0.51	0.68	0.22	0.71
C3	Manufacture of pharmaceuticals products, perfumes, soap and cleaning preparation	0.54	0.55	0.50	0.37	0.48	0.65	0.44	0.81
E3	Manufacture of electric and electronic equipment	0.49	0.42	0.38	0.35	0.54	0.68	0.35	0.71
G1	Extrac. of coal, crude petro., gas and uranium; man. of coke, ref. petrol. prod., and nuclear fuel	0.63	0.59	0.63	0.69	0.53	0.68	0.50	0.81

ABSORB: Average level of absorptive capacity
MKT: Capability to use the market for technologies
COOP: Competences of cooperation with other institutions
RD: Competence to manage a formal internal research activity
HET: Competences of stimulation of the individual initiative and creativity
HOM: Competences of knowledge sharing between workers
SCI: Competence to recruit highly qualified workers for the innovation
TRAIN: Competence to organize a continuous training of the employees

Table 20: Mean levels of competence of external interface in different industrial sectors (NES level 2) in the Competence survey over the period [1994-98]

NES level 2	Industry	MNGT	ACCU	EXPLOR	EVAL	IP
C1	Manufacture of clothing articles and leather products	0.33	0.40	0.41	0.27	0.23
C2	Publishing, printing and reproduction of recorded media	0.37	0.47	0.50	0.30	0.20
F1	Mining and quarrying except energy prod. materials, manuf. of other non-metallic mineral prod.	0.42	0.50	0.56	0.36	0.27
C4	Manufacture of domestic equipment	0.45	0.50	0.53	0.40	0.36
F3	Manufacture of wood, wood products, pulp, paper and paper products	0.42	0.54	0.54	0.38	0.23
F2	Manufacture of textiles	0.46	0.54	0.57	0.40	0.35
F5	Manufacture of basic metals and fabricated metal products	0.40	0.50	0.55	0.30	0.23
E2	Manufacture of metal products, machinery and equipment	0.43	0.50	0.54	0.38	0.30
E1	Building of ships and boats, manufacture of railway locomotives, rolling stock	0.50	0.55	0.63	0.49	0.33
F4	Manufacture of chemicals, rubber, plastic and chemical products	0.52	0.57	0.65	0.47	0.39
D0	Manufacture of motor vehicles	0.51	0.56	0.60	0.48	0.40
F6	Manufacture of electric and electronic components	0.52	0.57	0.68	0.43	0.41
C3	Manufacture of pharmaceuticals products, perfumes, soap and cleaning preparation	0.61	0.62	0.67	0.61	0.54
E3	Manufacture of electric and electronic equipment	0.55	0.58	0.66	0.53	0.41
G1	Extrac. of coal, crude petro., gas and uranium; man. of coke, ref. petrol. prod., and nuclear fuel	0.60	0.69	0.71	0.54	0.48

MNGT: Average level of competences for the management of the innovation

EXPLOR: Competence to manage an explorative innovation strategy

EVAL: Competence to manage a rational innovative strategy

ACCU: Competence to manage the accumulation of the current and past technological choices and activities

IP: Competence to manage the protection of the intellectual assets of the firm

APPENDIX 4 : RESULT FROM A PRINCIPAL COMPONENT ANALYSIS PERFORMED ON FIRMS' STANDARDIZED COMPETENCE SCORES

That principal component analysis was performed on the full set of firms that were included in the Competence survey, that is 3846 firms. 15 elementary competences were taken into account:

SUPP: Competences of external interface towards suppliers

USR: Competences of external interface towards customers

COMP: Competences of external interface towards competitors

WATCH: Competences of external interface towards the global technological environment

MKTTEC: Capability to use the market for technologies

COOP: Competences of cooperation with other institutions

RD: Competence to manage a formal internal research activity

HET: Competences of stimulation of the individual initiative and creativity

HOM: Competences of knowledge sharing between workers

SCI: Competence to recruit highly qualified workers for the innovation

TRAIN: Competence to organize a continuous training of the employees

EXPLOR: Competence to manage an explorative innovation strategy

EVAL: Competence to manage a rational innovative strategy

ACCU: Competence to manage the accumulation of the current and past technological choices and activities

IP: Competence to manage the protection of the intellectual assets of the firm

Beforehand firms' individual scores for each competence were standardized by their sector mean. Observations were weighted by survey weights.

Table 21: Eigenvalues of the correlation matrix

Principal componen t	Eigenvalues	Proportion of the total inertia
Axis 1	6.5645	43.8%
Axis 2	1.4203	9.5%
Axis 3	0.8382	5.6%
Axis 4	0.7902	5.3%
Axis 5	0.7242	4.8%
Axis 6	0.6527	4.4%
Axis 7	0.5876	3.9%
Axis 8	0.5260	3.5%
Axis 9	0.5006	3.3%
Axis 10	0.4991	3.3%
Axis 11	0.4697	3.1%
Axis 12	0.4266	2.8%
Axis 13	0.3867	2.6%
Axis 14	0.3260	2.2%
Axis 15	0.2878	1.9%

Materials for the interpretation are presented on the next page. We suggest here a synthetic interpretation.

Table 22: Coordinate of the competences on each principal component

Variable	PRIN1	PRIN2	PRIN3	PRIN4	PRIN5	PRIN6	PRIN7	PRIN8	PRIN9	PRIN10	PRIN11	PRIN12	PRIN13	PRIN14	PRIN15
SUPP	0.603	0.023	0.433	0.148	-0.011	0.536	-0.293	0.175	-0.073	0.087	0.085	0.010	0.046	0.026	0.000
USR	0.700	-0.094	-0.088	-0.429	-0.013	0.012	0.144	0.071	0.024	0.302	0.168	0.388	0.110	-0.014	0.028
COMP	0.751	0.022	0.050	-0.218	-0.037	-0.027	0.198	-0.004	-0.246	0.161	0.239	-0.375	-0.244	-0.019	-0.018
WATCH	0.539	0.249	0.465	0.019	0.545	-0.077	0.296	-0.128	0.041	-0.080	-0.108	0.057	0.014	0.012	0.039
COOP	0.542	0.574	-0.158	0.151	-0.078	-0.009	-0.101	-0.258	0.046	0.408	-0.235	-0.093	0.095	0.065	-0.011
RD	0.579	0.444	-0.110	-0.087	0.210	-0.268	-0.464	-0.023	0.023	-0.168	0.278	0.060	-0.071	0.015	0.031
MKTTEC	0.462	0.538	-0.369	0.246	-0.085	0.313	0.326	0.058	0.022	-0.234	0.144	0.089	-0.028	-0.024	0.008
HET	0.673	-0.410	-0.079	0.309	-0.048	-0.071	-0.004	-0.195	-0.291	-0.038	-0.012	0.174	-0.108	0.283	0.154
HOM	0.782	-0.309	-0.102	0.141	0.025	0.017	-0.064	-0.113	0.057	0.000	-0.050	-0.042	0.013	-0.413	0.257
SCI	0.545	0.181	0.432	0.195	-0.533	-0.346	0.094	0.075	0.114	-0.050	0.062	0.075	0.011	-0.046	-0.014
TRAIN	0.633	-0.247	-0.192	0.352	0.247	-0.187	0.038	0.445	0.191	0.170	0.003	-0.088	0.006	0.074	-0.063
EXPLOR	0.789	-0.307	-0.052	0.093	0.033	0.023	-0.038	-0.243	-0.019	-0.079	0.015	0.063	0.021	-0.119	-0.429
ACCU	0.739	-0.246	-0.034	-0.197	-0.083	0.077	0.030	-0.144	0.294	-0.178	0.095	-0.241	0.295	0.205	0.074
EVAL	0.746	-0.023	-0.031	-0.280	-0.101	0.115	-0.069	0.071	0.246	-0.108	-0.330	0.053	-0.376	0.065	-0.003
IP	0.726	0.139	-0.082	-0.204	-0.044	-0.101	-0.031	0.255	-0.380	-0.189	-0.295	-0.046	0.238	-0.034	-0.020

Table 23: Contribution of the competences to each principal component

Variables	PRIN1	PRIN2	PRIN3	PRIN4	PRIN5	PRIN6	PRIN7	PRIN8	PRIN9	PRIN10	PRIN11	PRIN12	PRIN13	PRIN14	PRIN15
SUPP	5.5%	0.0%	22.3%	2.8%	0.0%	43.9%	14.6%	5.8%	1.1%	1.5%	1.5%	0.0%	0.6%	0.2%	0.0%
USR	7.5%	0.6%	0.9%	23.3%	0.0%	0.0%	3.5%	1.0%	0.1%	18.3%	6.0%	35.3%	3.1%	0.1%	0.3%
COMP	8.6%	0.0%	0.3%	6.0%	0.2%	0.1%	6.7%	0.0%	12.1%	5.2%	12.2%	32.9%	15.4%	0.1%	0.1%
WATCH	4.4%	4.4%	25.8%	0.0%	41.0%	0.9%	14.9%	3.1%	0.3%	1.3%	2.5%	0.8%	0.1%	0.0%	0.5%
COOP	4.5%	23.2%	3.0%	2.9%	0.8%	0.0%	1.7%	12.7%	0.4%	33.3%	11.8%	2.0%	2.3%	1.3%	0.0%
RD	5.1%	13.9%	1.4%	1.0%	6.1%	11.0%	36.6%	0.1%	0.1%	5.7%	16.4%	0.8%	1.3%	0.1%	0.3%
MKTYEC	3.3%	20.4%	16.3%	7.7%	1.0%	15.0%	18.1%	0.6%	0.1%	10.9%	4.4%	1.9%	0.2%	0.2%	0.0%
HET	6.9%	11.8%	0.7%	12.1%	0.3%	0.8%	0.0%	7.2%	16.9%	0.3%	0.0%	7.1%	3.0%	24.5%	8.3%
HOM	9.3%	6.7%	1.2%	2.5%	0.1%	0.0%	0.7%	2.4%	0.6%	0.0%	0.5%	0.4%	0.0%	52.4%	22.9%
SCI	4.5%	2.3%	22.3%	4.8%	39.2%	18.3%	1.5%	1.1%	2.6%	0.5%	0.8%	1.3%	0.0%	0.7%	0.1%
TRAIN	6.1%	4.3%	4.4%	15.7%	8.4%	5.4%	0.2%	37.6%	7.2%	5.8%	0.0%	1.8%	0.0%	1.7%	1.4%
EXPLOR	9.5%	6.7%	0.3%	1.1%	0.2%	0.1%	0.2%	11.2%	0.1%	1.3%	0.0%	0.9%	0.1%	4.3%	64.0%
ACCU	8.3%	4.3%	0.1%	4.9%	0.9%	0.9%	0.2%	3.9%	17.2%	6.4%	1.9%	13.6%	22.6%	12.9%	1.9%
EVAL	8.5%	0.0%	0.1%	9.9%	1.4%	2.0%	0.8%	1.0%	12.1%	2.3%	23.2%	0.7%	36.6%	1.3%	0.0%
IP	8.0%	1.4%	0.8%	5.3%	0.3%	1.6%	0.2%	12.3%	28.9%	7.1%	18.6%	0.5%	14.6%	0.4%	0.1%

Table 24: Cos² of the competences on each principal component

Variable	PRIN1	PRIN2	PRIN3	PRIN4	PRIN	PRIN6	PRIN7	PRIN8	PRIN9	PRIN10	PRIN11	PRIN12	PRIN13	PRIN14	PRIN15
SUPP	36.4%	0.1%	18.7%	2.2%	0.0%	28.7%	8.6%	3.1%	0.5%	0.8%	0.7%	0.0%	0.2%	0.1%	0.0%
USR	49.0%	0.9%	0.8%	18.4%	0.0%	0.0%	2.1%	0.5%	0.1%	9.1%	2.8%	15.0%	1.2%	0.0%	0.1%
COMP	56.3%	0.0%	0.3%	4.8%	0.1%	0.1%	3.9%	0.0%	6.1%	2.6%	5.7%	14.1%	6.0%	0.0%	0.0%
WATCH	29.0%	6.2%	21.6%	0.0%	29.7%	0.6%	8.7%	1.6%	0.2%	0.6%	1.2%	0.3%	0.0%	0.0%	0.2%
COOP	29.4%	33.0%	2.5%	2.3%	0.6%	0.0%	1.0%	6.7%	0.2%	16.6%	5.5%	0.9%	0.9%	0.4%	0.0%
RD	33.5%	19.7%	1.2%	0.8%	4.4%	7.2%	21.5%	0.1%	0.1%	2.8%	7.7%	0.4%	0.5%	0.0%	0.1%
MKTTEC	21.4%	29.0%	13.6%	6.1%	0.7%	9.8%	10.6%	0.3%	0.0%	5.5%	2.1%	0.8%	0.1%	0.1%	0.0%
HET	45.3%	16.8%	0.6%	9.6%	0.2%	0.5%	0.0%	3.8%	8.5%	0.1%	0.0%	3.0%	1.2%	8.0%	2.4%
HOM	61.2%	9.5%	1.0%	2.0%	0.1%	0.0%	0.4%	1.3%	0.3%	0.0%	0.2%	0.2%	0.0%	17.1%	6.6%
SCI	29.7%	3.3%	18.7%	3.8%	28.4%	12.0%	0.9%	0.6%	1.3%	0.3%	0.4%	0.6%	0.0%	0.2%	0.0%
TRAIN	40.1%	6.1%	3.7%	12.4%	6.1%	3.5%	0.1%	19.8%	3.6%	2.9%	0.0%	0.8%	0.0%	0.5%	0.4%
EXPLOR	62.3%	9.4%	0.3%	0.9%	0.1%	0.1%	0.1%	5.9%	0.0%	0.6%	0.0%	0.4%	0.0%	1.4%	18.4%
ACCU	54.5%	6.0%	0.1%	3.9%	0.7%	0.6%	0.1%	2.1%	8.6%	3.2%	0.9%	5.8%	8.7%	4.2%	0.5%
EVAL	55.7%	0.1%	0.1%	7.9%	1.0%	1.3%	0.5%	0.5%	6.0%	1.2%	10.9%	0.3%	14.2%	0.4%	0.0%
IP	52.7%	1.9%	0.7%	4.2%	0.2%	1.0%	0.1%	6.5%	14.5%	3.6%	8.7%	0.2%	5.6%	0.1%	0.0%

REFERENCES

Audretsch, D. B. (1995), "Firm profitability, growth and innovation", *Review of Industrial Organization*, no.10, pp.579-588.

Cefis, E. (2003), "Is there persistence in innovative activities?," *International Journal of Industrial Organization*, 21(4), April, pp.489-515

Cefis, E. (1999), "Persistence in profitability and in innovative activities", European Meeting on Applied Evolutionary Economics, 7-9 June 1999, Grenoble, France.

Cefis, E. and Orsenigo, L. (2001), "The persistence of innovative activities: A cross-countries and cross-sectors comparative analysis", *Research Policy*, 30(7), August, pp.1139-1158.

Cohen, W. M. and Levinthal, D. A. (1989), "Innovation and learning: The two faces of R&D", *The Economic Journal*, 99, September, pp.569-596.

Foray, D. and Mairesse, J. (1999), *Innovations et Performances. Approches interdisciplinaires*, édition des Hautes Études en Sciences Sociales, Paris

François, J-P. (1998), "Les compétences pour innover," Le 4 Pages, SESSI, n°85, janvier

François, J-P. (1998), "Les compétences pour innover dans l'industrie," Chiffres Clés Référence, SESSI

François, J-P., Goux, D., Guellec, D. and Templé, P. (1999), "Décrire les compétences pour l'innovation

Une proposition d'enquête", in Innovations et Performances. Approches Interdisciplinaires, ed. by

Foray, D. and Mairesse, J., édition des Hautes Études en Sciences Sociales, Paris, pp.283-303

Leiponen, A. (1997), "Dynamic competences and firm performance", IIASA, Interim Report n°IR-97-006/February

Leiponen, A. (2000), Competencies, Innovation and Profitability of Firms, *Economics of Innovation and New Technology*, 9(1), pp.1-24

Malerba, F. (1992), "Learning by firms and incremental technical change", *The Economic Journal*, vol.102, July, pp.845-859.

Malerba, F. and Orsenigo, L. (1993), "Technological regimes and firm behavior", *Industrial and Corporate Change,* 2(1), pp.45-71.

Malerba, F., Orsenigo, L. and Peretto, P. (1997), "Persistence of innovative activities, sectoral patterns of innovation and international technological specialization", *International Journal of Industrial Organization*, 15(6), October, pp.801-826

Nelson, R. R. (1994), "The role of firm difference in an evolutionary theory of technical advance", in *Evolutionary and Neo-Schumpeterian Approaches to Economics*, ed. by Magnusson L., Kluwer Academic Publishers, Norwell, Massachusetts, USA, pp.231-242.

Nelson, R. R. and Winter, S. G. (1982), *An Evolutionary Theory of Economic Change*, The Belknap Press of Harvard University Press, London

Waring, G. F. (1996), "Industry differences in the persistence of firm-specific returns", *The American Economic Review*, 86(5), December, pp.1253-1265.

Chapter 4

CHARACTERISTICS OF PERSISTENT INVENTORS AS REVEALED IN PATENT DATA

William Latham, *University of Delaware*
Christian Le Bas, *University of Lyon 2*
Karim Touach, *University of Lyon 2*

1. INTRODUCTION: THE PROLIFIC INVENTORS

There is empirical evidence regarding the role of certain key individuals in the creation of technological knowledge. In particular, studies of knowledge-based industries have identified the significant roles of key scientists and key engineers in the process of technological innovation (Zucker and Darby, 1996, 1998, and Almeida and Kogut, 1997). With respect to the process of technological invention, Gay (2003) using patent data provided by the US Patent Office found a high positive correlation between the presence of *consistent inventors* or *strong inventors* (who patented from 10 to 50 times during the period under observation) and the technological value of patented inventions as measured by the number of citations to the patent by subsequent patents. The Patent Value survey provides the opportunity to make value comparisons. Rothwell (1992) pointed out that, among the factors which could explain the success of a firm's innovative strategy, was the presence of certain key individuals: effective "product champions" or "technological gatekeepers". He considered product champions and project leaders as playing an important role in achieving both more successful and faster new product development. Many other scholars in the field of management and organization have emphasized the importance of individuals in the generation and promotion of ideas in the innovation process as well. Individuals emerge to actively promote innovations through the various organizational stages. Theses individuals are pivotal to successful implementation of an innovation. Case studies of innovation highlight the role of these champions (Howell and Boies, 2004). The champions (1) overcome the social pressures that an organization may impose in opposition to innovation, (2) demonstrate personal commitment to a new idea, (3) "promote the idea with conviction, persistence, and energy through informal networks," and (4) "willingly risk their position and reputation to ensure its success"

(Howell and Boies, 2004)[1]. All these studies, as well as many others, indicate clearly that the importance of individuals in the process of new knowledge creation has become more and more clearly recognized. This evolution now compels us to consider human capital, or even talent, as a real factor of growth (Florida, 2001).

Recent studies have revealed that these star scientists and key engineers are important, not only in the process of technological innovation, but also in the production of value (Zucker and Darby, 1996, 1998, Almeida and Kogut, 1997, and Stople, 2001). Numerous papers show that the roles played by key individuals are the result of both their knowledge, which is largely tacit and specialized, and their entrepreneurial capacities (Mangematin *et al.*, 2003). These scientific and technological elites are active in both the production and diffusion of knowledge and in the localization of technological success. From patent data provided by A. B. Jaffe and M. Trajtenberg, C. Gay (2003) shows that these very active individuals, her "strong and star inventors", are, in fact, participants in many inventions[2]. She verifies a positive and high correlation between the presence of "strong and star inventors"[3] and the technological value of patented inventions as measured by the number of subsequent citations received by a patent. Latham *et al.* (2005) have analyzed U.S. National Bureau of Economic Research (NBER) data on individual inventors named in patents for three countries, France, Germany and the UK. They provide significant evidence of a positive relationship between the presence on the inventive team of a prolific inventor, defined as an individual named in ten or more patent applications, and the size of this team. They also find a positive relationship between the presence of a prolific inventor and the value of an invention as measured by the number of citations it receives in subsequent patents. This result supports a concept developed principally in bibliometric analyses: a small number of prolific authors produces a large part of the knowledge of greatest value. These previous empirical studies have documented the crucial importance of these prolific inventors, who invent consistently, and who produce high value innovations. We find these studies to be intriguing and they motivate us to learn more about this central figure in the inventive process, the prolific inventor.

In this chapter we will emphasize a dimension that has not been dealt with previously: a prolific inventor is likely to be a persistent inventor, an inventor who remains active over a long period of time. We retain the definition of a prolific inventor as one who is named in at least ten patents applications over a long period of time (Gay, 2003, Latham et al., 2005).

This chapter is structured as follows. In the next section we describe our patent data and define the variables we will use in our analysis. In section 3 we present some descriptive analysis regard the general scale and scope of the population of prolific inventors found in our sample. Then, in section 4, the characteristics of prolific inventors are identified through logistic regression analysis. Finally, in section 5, we explore more rigorously the relationship between consistence and persistence in the invention process at the level of individual prolific inventors.

2. DESCRIPTION OF DATA AND VARIABLES

The empirical analysis of this chapter uses patent data to draw conclusions about the innovation process. Although patent data have well-known drawbacks for this purpose[4], patent statistics are a unique resource for the analysis of the process of technical change (Griliches, 1991). It is possible to use patent data to assess the patterns of innovation activity, especially the variations in its rhythms and intensity across firms, technological fields, nations, regions, etc. Following the seminal work of J. Schmookler (1966), it is now common to use data that come from U.S. patents granted by the United States Patent and Trademark Office (USPTO) rather than the patents of other nations in economic studies dealing with technological accumulation. We use NBER patent data files to obtain detailed information on U.S. patents granted between 1975 and 1999 as well as information on the inventors (for a detailed description of these data, see Hall *et al.*, 2001, and Jaffe and Trajtenberg, 2002). We only deal with patents originating in France, Germany, the United Kingdom and Japan, *i.e.*, patents whose primary inventors are French, German, British or Japanese nationals)[5]. A patent document contains considerable information concerning the process of knowledge creation. Each patent granted by the U.S. patent office provides the title and description of the invention, the names and addresses of the inventor(s) and assignee(s), citations made to other patents and to scientific papers, etc. Recently these patent data have become available in computer-readable or directly-downloadable forms. This publicly available information contained in patents offers perspectives on the process of innovation unobtainable from traditional patent count data. In this chapter the phenomena of interest are the following[6]:

– The names of all inventors. Patent data provide, for each patent, the names of all inventors. The occurrence of the same inventor's name on multiple patents in the data base permits us to determine which ones are the prolific inventors (our key individuals) and which ones are not. We identify as "prolific" those inventors whose names appeared on the list of inventors in patent applications for more than 10 patents during the period of the data (24 years).

– The numbers of inventors listed on each patent. The number of inventors listed in the patent document is assumed to be highly correlated with the size of the inventive team which performed the research leading to the patent. The size of the research team is interpretable as an indicator of the collective dimension of the knowledge creation. We use this indicator in the same way that it is used in bibliometrics: there, the number of authors gives an indication of collective creation in scientific publications. Although we have evidence that the actual number of people who participate in the creation of new ideas is larger than the number of inventors listed in a patent document, we can still use the number of inventors as a proxy for the scale of collective invention. Obviously, patent data do not specify the kinds of links that may exist between and among the listed inventors. We assume that all co-inventors have collaborated in the creation of new knowledge.

– The nationality of all the inventors listed. By knowing the nationalities of the individual inventors, we are able to identify, in particular, the involvement of a foreign inventor, that is an inventor with a nationality different from that of the primary inventor.

– The technological category of the patent. Each patent is classified as contributing primarily to a specific, detailed technological category. The detailed categories can be aggregated to broader categories. We use the most aggregate level, which has only six broad technology categories:

– Chemicals,

– Computers and Communications,

– Drugs and Medical,

– Electrical and Electronic,

– Mechanical, and

– Other.

– The year in which each patent "issues," which is the year when the patent rights are granted. In the U.S. data, a patent usually issues approximately 18 month after the application date.

3. WHO ARE THE PROLIFIC INVENTORS? SOME DESCRIPTIVE STATISTICS

The total number of patents in the data base is 704,154. The total number of inventors named on the patents is 438,575 and the total number of prolific inventors is 36,600. The percentage distributions of numbers of patents per inventor for each country are shown in Table 1 for each of the four countries in the analysis. Table 1 also shows the percentage distributions of patents with prolific inventors by country. The percentage is between 4.36% and 7.41% for the three European countries, but is significantly higher at 10.64% for Japan. Table 1 also displays the proportion of very prolific inventors (the "star" inventors who are named in fifty patents and more) as well. The countries' differences are clear and significant. Table 2 shows that, when the number of prolific inventors is low in relation to the overall number of inventors, the number of patents that are related to the prolific inventors is much more important (see Table 2). In order to take into account the overall level of patenting by countries we have calculated the proportion of patents in which a prolific inventor is present (Table 2). The country differences are again striking. France and the UK have nearly the same proportions, 40.45% and 42.29% respectively. For Germany this indicator is significantly higher at 66.93%, but, for Japan it rises all the way to 94.41%. This last figure indicates that, for Japan, a prolific inventor is involved either individually or as a member of the inventive team, for nearly all Japanese patents[7].

Table 1. Percentage distribution of inventors by number of patents by country

Number of patents ⇓	France	Germany	UK	Japan	TOTAL
1	56.72	50.24	52.14	42.74	47.32
2-9	38.76	41.83	42.03	45.89	43.59
10-49	4.36	7.41	5.60	10.64	8.53
>49	0.16	0.53	0.23	0.73	0.56
TOTAL	100.00	100.00	100.00	100.00	100.00

Table 2. Number and proportion of patents including a prolific inventor by country

	France	Germany	UK	Japan
Total number of patents in the data set	65,018	65,038	169,849	390,769
Number of patents including a prolific inventor	26,299	111,985	24,448	368,844
Proportion of patents including a prolific inventor	40.45%	65.93%	42.29%	94.41%

4. A LOGISTIC REGRESSION FOR PROLIFIC INVENTORS CHARACTERISTICS

We use a logistic regression to study the effects of our variables on the probability of an inventor being prolific. In this regression, we use a sample of 438,575 patentees and have several different characteristics of the inventors related to whether they are prolific or not. Three independent variables are used: The inventor's ability to function as a member of a team. We hypothesize that an inventor who usually works more collectively is more likely to be a prolific inventor.

1) The inventor's country of residence implicitly informs us about, or is a proxy for, the national system of innovation of the nation in which the inventor works. We hypothesize that some characteristics of a nation's system of research and higher education push its inventors to be prolific. We have seen in the previous section that Japan has a much higher proportion of prolific inventors.

2) The inventor's field of technological competence, proxied by the technological field in which a patent applies. We hypothesize that different technological fields will produce different proportions of prolific inventors.

The variables used are summarized in Table 3 below:

Table 3. Variables used in the regression

Variable Description	Short Name	Type Qualitative (0 – 1)
Dependent Variable		
Whether the inventor is prolific or not	PROLIFIC	Qualitative (0 – 1)
Independent Variables		
The average number of inventors named in patents where our inventor is named	AVENUM	Quantitative (#)
Country of the inventor : France Great Britain Germany Japan	FR GB DE JP	Qualitative (0 – 1) Qualitative (0 – 1) Qualitative (0 – 1) Qualitative (0 – 1)
Primary technological category of invention Chemicals Computers and communications Drugs and medical Electrical and electronic Mechanical Other	CHEM COMP DRUG ELEC MECH OTHR	Qualitative (0 – 1) Qualitative (0 – 1) Qualitative (0 – 1) Qualitative (0 – 1) Qualitative (0 – 1) Qualitative (0 – 1)

The model that expresses our basic hypotheses is:

PROLIFIC = f (β_1 AVENUM + β_2 FR + β_3 GB + β_4 DE + β_5 JP + β_6 CHEM + β_7 COMP + β_8 DRUG+ β_9 ELEC+ β_{10} MECH+ β_{11} OTHR)

In this model, we attempt to explain the probability that an inventor is prolific by using (1) his ability to work as part of a team (AVENUM), (2) his own country, and (3) his field of technological competence. The descriptive statistics concerning these variables are given in the Appendix to this chapter. We want to obtain coefficients for the dummy variables that are interpretable as they commonly are as the difference it makes in the value of the dependent variable to be in a part of the data where an included category has a value of 1 in comparison with being in a part of the data where an excluded category has a value of 1. Thus we will add a constant term to the equation and omit one category for each dummy variable. GB is the omitted country and CHEM is the omitted technological field. We also want to be able to examine whether or not there are significant interactions among the variables. For example, we want to know if the effects of AVENUM are different by country or by technological field. Thus we will also add to the equation all of the interactive terms, 3 for the included countries with AVENUM, 5 for the included technological fields with AVENUM, and 20 for the interactions between the included dummies.

PROLIFIC = f ($\beta 1$ AVENUM + $\beta 2$ COMP + $\beta 3$ DRUG + $\beta 4$ ELEC + $\beta 5$ MECH + $\beta 6$ OTHR + $\beta 7$ FR + $\beta 8$ JP + $\beta 9$ DE + $\beta 10$ COMP*FR + $\beta 11$ DRUG*FR + $\beta 12$ ELEC*FR + $\beta 13$ MECH*FR + $\beta 14$ OTHR*FR + $\beta 15$ COMP*JP + $\beta 16$ DRUG*JP + $\beta 17$ ELEC*JP + $\beta 18$ MECH*JP + $\beta 19$ OTHR*JP + $\beta 20$ COMP*DE + $\beta 21$ DRUG*DE + $\beta 22$ ELEC*DE + $\beta 23$ MECH*DE + $\beta 24$ OTHR*DE + $\beta 25$ COMP*AVENUM + $\beta 26$ DRUG*AVENUM + $\beta 27$ ELEC*AVENUM + $\beta 28$ MECH*AVENUM + $\beta 29$ OTHR*AVENUM + $\beta 30$ FR*AVENUM + $\beta 31$ JP*AVENUM + $\beta 32$ DE*AVENUM + C)

The dependent variable is binary and the regression will be based on the specification of a logit function. The β's are estimated with the maximum likelihood method and robust covariances are produced using the GLM procedure. Alternative specifications and models were tested, including a series of Probit models. The full interaction logit model was selected based on AIC and McFadden R-square statistics.

When models for each country were estimated individually, all the independent variables were highly significant. However, models using country dummy variables are preferred, since they have higher

R-squares and their LR tests are all significant.

Models with each technological fields estimated individually do not perform as well, and equations for only two of the individual technological fields would converge.

Models without a constant term had similar results (similar R-square and AIC), compared to models with a constant.

The most general model with all the interaction terms and all variables was tested with Wald's restriction tests. Most of the interaction terms with the selected model are significant at 10% level and many at higher levels of significance.

The models generally have very good prediction probabilities. However, goodness-of –fit tests are not as good. Because AVENUM is highly skewed, heteroscedascticity tests were done by running a regression using the standardized residual as dependent variables. While the results indicate some heteroscedasticity, with so many observations, we do not believe that it is too much of a problem.

Marginal effects were computed (but only for the major variables and major interactions) using the predicted index values as marginal effect. The listed effects in the appendix table are mean marginal effects. They are mostly quite small, which is not surprising. The largest shows that becoming a German inventor only increases the probability of being a prolific inventor by less than 1%.

The results of estimating the equation are shown in Table 4. Overall the regression explains most of the variation in whether or not an inventor becomes prolific, the McFadden R-square is .83. All of the independent variables alone are significant confirming our hypotheses about their effects. The principal result is the positive and significant effect of the inventor's propensity to work as a member of a team, once we control for technological fields and country effects.

There are interesting results among the interaction terms. The interactions of technological fields with countries are mostly not significant but for ELEC the effect is negative in both France and Japan. On the other hand DRUG is positive in Japan. For Germany, all the included technological fields have negative effects. Interactions with AVENUM are most significant for Japan and for DRUG, MECH and OTHR. These results indicate that the process of becoming prolific does vary across countries and technological fields and that the effects of teams of inventors also vary in these dimensions. These results provide support for the idea that individual country institutions are important in the innovative process.

Table 4a. Estimation results: Equation Statistics

Variable	Coefficient	Std. Error	z-Statistic	Prob.
Dependent Variable: PROLIFIC				
Method: MAXIMUM LIKELIHOOD - Binary Logit (Quadratic hill climbing)				
Included observations: 438575				
C	-6.946820	0.133732	-51.94592	0.0000
AVENUM	0.171654	0.028556	6.011188	0.0000
COMP	0.372319	0.034305	10.85308	0.0000
DRUG	0.687237	0.026832	25.61222	0.0000
ELEC	0.471965	0.036736	12.84762	0.0000
MECH	0.429997	0.033075	13.00062	0.0000
OTHR	0.474186	0.033118	14.31787	0.0000
FR	0.466251	0.183034	2.547342	0.0109
JP	0.589029	0.147056	4.005481	0.0001
DE	1.219169	0.143907	8.471945	0.0000
COMP*FR	0.002941	0.039073	0.075270	0.9400
DRUG*FR	-0.051385	0.033456	-1.535910	0.1246
ELEC*FR	-0.145882	0.045232	-3.225164	0.0013
MECH*FR	-0.010993	0.040274	-0.272949	0.7849
OTHR*FR	-0.001464	0.040209	-0.036400	0.9710
COMP*JP	0.042011	0.033345	1.259899	0.2077
DRUG*JP	0.065734	0.027555	2.385537	0.0171
ELEC*JP	-0.062447	0.036951	-1.689983	0.0910
MECH*JP	-0.067191	0.033353	-2.014543	0.0440
OTHR*JP	-0.011878	0.033055	-0.359332	0.7193
COMP*DE	-0.086110	0.037411	-2.301738	0.0213
DRUG*DE	-0.047609	0.027640	-1.722463	0.0850
ELEC*DE	-0.190388	0.037902	-5.023225	0.0000
MECH*DE	-0.105269	0.033286	-3.162559	0.0016
OTHR*DE	-0.086640	0.032987	-2.626501	0.0086
COMP*AVENUM	0.000110	0.004284	0.025746	0.9795
DRUG*AVENUM	-0.025989	0.003419	-7.601043	0.0000
ELEC*AVENUM	0.001186	0.003876	0.305903	0.7597
MECH*AVENUM	0.019725	0.004042	4.880362	0.0000
OTHR*AVENUM	0.046014	0.004935	9.324334	0.0000
FR*AVENUM	-0.025222	0.042583	-0.592303	0.5536
JP*AVENUM	-0.111935	0.030406	-3.681360	0.0002
DE*AVENUM	0.020809	0.030191	0.689232	0.4907

Table 4b. Estimation results: Summary Statistics

Mean dependent var	0.083452	S.D. dependent var	0.276565
S.E. of regression	0.095803	Akaike info criterion	0.099497
Sum squared resid	4025.005	Schwarz criterion	0.100324
Log likelihood	-21785.44	Hannan-Quinn criter.	0.099732
Restr. log likelihood	-125923.9	Avg. log likelihood	-0.049673
LR statistic (32 df)	208277.0	McFadden R-squared	0.826995
Probability(LR stat)	0.000000		
Obs with Dep=0	401975	Total obs	438575
Obs with Dep=1	36600		

5. THE PROLIFIC INVENTOR AS A PERSISTENT INVENTOR

We now want to explore the temporal dimension of being prolific (or consistence). A prolific inventor is usually associated with patent applications submitted during different periods of time (at least for a great proportion of prolific inventors). In order to study this phenomenon we constructed a random sample of 337 prolific inventors. Working on the entire data set with the huge number of inventors it contains was, at this stage of the research, prohibitively expensive in terms of time and other resources. Amongst this randomly drawn population of inventors there are 220 inventors whose inventions all appear during a single spell, whatever its length, and 100 whose inventions appear in two spells. A small minority (17) invent sporadically, with more than 2 spells[8]. We begin by exploring the relationship between consistence and persistence as revealed in the patent documents at the inventor level. In Table 5 we present the frequency distributions for both the total number of spells and the maximum length of spells (the longest spell for each inventor) for our sample of 337 prolific inventors observed over the period from 1975 to 1999. With respect to the distribution of all spells, the most notable feature is that 59 spells (12.53 %) are very short, one-year spells. As the spell length increases, the number of spells begin to increase (irregularly) until 9 years, and then decreases (irregularly as well). There is a small spike for the longest spell at 25 years due to truncation of the data at this length. As regards the maximum length of spells, the features of the distribution are more stable. The distribution is not really symmetrical because the mode of the distribution is 9 years.

Table 5. Distribution of patenting spell lengths and maximum spell lengths for 337 inventors

Spell Length (Years)	All spells			Maximum spells		
	No.	%	Cum. %	No.	%	Cum. %
1	59	12.53	12.53	0	0.00	0.00
2	20	4.25	16.77	2	0.59	0.59
3	22	4.67	21.44	6	1.78	2.37
4	26	5.52	26.96	8	2.37	4.75
5	22	4.67	31.63	13	3.86	8.61
6	17	3.61	35.24	10	2.97	11.57
7	32	6.79	42.04	27	8.01	19.58
8	26	5.52	47.56	25	7.42	27.00
9	41	8.70	56.26	40	11.87	38.87
10	27	5.73	62.00	27	8.01	46.88
11	30	6.37	68.37	30	8.90	55.79
12	20	4.25	72.61	20	5.93	61.72
13	14	2.97	75.58	14	4.15	65.88
14	16	3.40	78.98	16	4.75	70.62
15	16	3.40	82.38	16	4.75	75.37
16	15	3.18	85.56	15	4.45	79.82
17	8	1.70	87.26	8	2.37	82.20
18	15	3.18	90.45	15	4.45	86.65
19	4	0.85	91.30	4	1.19	87.83
20	7	1.49	92.78	7	2.08	89.91
21	7	1.49	94.27	7	2.08	91.99
22	3	0.64	94.90	3	0.89	92.88
23	4	0.85	95.75	4	1.19	94.07
24	5	1.06	96.82	5	1.48	95.55
25	15	3.18	100.00	15	4.45	100.00
Total	471	100.00		337	100.00	

Next we report the results obtained from statistical exploration of the relationship between consistence and persistence in our sample of 337 inventors. First we regressed the total number of patents obtained by a each inventor on the spell length for all the spells and on the length of the maximum spell. In a log-log model we find statistically significant relationships (as indicated by R-square) for both the all spells and the maximum spells models. The relationship is more significant when we consider all the spells. The results were as expected: the more persistently an inventor is, the more he invents.

In order to shed some additional light on this relationship, we attempted to test for the existence of a link between the number of patents granted at the beginning of the spell and the time duration of it. The works of Geroski *et al.* (1997) and the findings of Chapter 2 of this volume corroborate this hypothesis for *firm* innovations. The results from our regressions for individual inventors (see Appendix 2) show no evidence confirming these firm results. One of our regressions even indicates that the reverse might be true: that the persistent inventor is not necessarily a consistent one at the outset of his or her invention period. This result may be interpreted as showing that the same dynamics is not at work for individuals as is present for firms. In general, a single firm owns a research team that may include a number of individuals who are engaged in research and inventing on the firm's behalf while they remain with the firms, but who are mobile. A prolific inventor does not necessarily stay in the same enterprise during for his or her entire inventing life.

While the link between the number of patents and spell length may be obvious, the link between spell length and inventive performance is not. By inventive (or technological) performance we mean the inventor's capacity to produce new ideas. We measure it by the ratio of the number of patents in a spell to the spell length (in years). In a sense it measures the inventor's productivity. The model that provides the best fit is the following quadratic equation:

Inventive performance = $\beta_1 + \beta_2 *$ (spell length) + $\beta_3 *$ (spell length)^2

We find (Student's t-values are in parentheses): R-square = 0.18, β_3 = 0.013 (8.510), $\beta_2 = -0.365$ (- 8.406), $\beta_1 = 3.932$ (14.026) with n = 337. The F-value for the test for overall significance of the regression is highly significant (at the 1% level of significance). This U-shaped relationship is not easy to interpret. It seems to indicate that there are decreasing returns with regard to duration for the shortest spell lengths but that the reverse becomes true for the inventors having the longest spells (after 15 years, precisely). For them, there is a positive relationship between their technological performance and the duration of their inventive spell. If we isolate the most prolific inventors, we find they are located on the growing part of the parabola where "dynamic economies of scale" or "learning economies" play a significant role. The longer is the spell, the higher is the inventive performance.

6. CONCLUSION

In this chapter we have provided empirical analyses regarding persistent and consistent inventors and we have provided evidence regarding some inventor characteristics which help to explain why an inventor becomes prolific. The underlying idea is that the prolific inventor supports the firm's accumulation of knowledge (technological learning) which is at the core of firm innovation persistence. However, prolific inventors and firm innovative persistence are not simply two faces of the same coin because innovative persistence could result from two different models of the firm innovative process. In the first, the firm's persistence in innovation is attributable to the inventive capacity of its team of inventors who are mostly permanent individuals. Japanese firms in the 1970s and 1980s provide good examples of this type of firm innovation. In the second model more mobile. In this model the inventors may be persistent and prolific but not in the same organization. The firm's capacity for innovation would be not affected only if the individual and collective capacities to learn were high. An extreme case could occur if inventors stayed only a few years in any firm. We think that the great majority of cases are intermediate: inventors do move from firm to firm, but not all inventors move and those who do move, do not do so often. Thus it may be that there is no connection between firm innovation persistence and persistence of the production of new ideas at the individual (inventor) level. The crucial phenomenon is the inventor mobility. A future research task might be to assess the scale and scope of the mobility and its effects on firm innovative performance and persistence.

ENDNOTES

1 Champions distinguish themselves from non-champions. It is by "communicating a clear vision of what innovation could be, that champions play a decisive role in implementing new ideas by communicating strategic meaning around the innovations... "(Howell and Boies, 2004). Champions must have a broad knowledge and vision of their roles (Mumford et al., 2002) and must work collaboratively with people. From the cases studies carried out by Howell and Boies (2004) it appears clear that champions demonstrate more enthusiastic support for new ideas, tie innovations to a greater variety of positive organizational outcomes, and use informal selling more often during idea promotion.

2 In her study a "strong inventor" is named in at least ten patent applications and a "star inventor" is named in at least fifty.

3 This notion is closely related to the "star scientists" of Zucker and Darby (1994). We use the term "prolific inventor" hereafter in this paper.

4 See those described in detail in Chapter 2 of this volume.

5 In comparison with Gay et al., 2005, we have increased our sample size significantly by including Japanese patents.

6 Other crucial information as the value of patent (given by the amount of citations received) might be useful for the next step of the research.

7 Note that more than one prolific inventor can be named in any single patent document.

8 With respect to the spell length, we consider a spell to have ended if three years elapse without an invention. This means that two spells must be separated by at least three years.

APPENDIX 1 : DESCRIPTIVE STATISTICS OF THE DATA USED IN LOGIT MODEL

	AVERAGE INVENTORS	FRANCE	GB	GERMANY	JAPAN	TECH-FIELD 2	TECH-FIELD 1	TECH-FIELD 3	TECH-FIELD 4	TECH-FIELD 5	TECH-FIELD 6
Mean	3.357515	0.118016	0.101007	0.260710	0.523915	0.725557	0.991016	0.658567	1.607978	2.402996	1.352079
Median	3.000000	0.000000	0.000000	0.000000	1.000000	0.000000	0.000000	0.000000	0.000000	0.000000	0.000000
Maximum	34.00000	1.000000	1.000000	1.000000	1.000000	266.0000	364.0000	918.0000	1436.000	1105.000	1542.000
Minimum	1.000000	0.000000	0.000000	0.000000	0.000000	0.000000	0.000000	0.000000	0.000000	0.000000	0.000000
Std. Dev.	1.905744	0.322628	0.301338	0.439023	0.499428	3.923003	3.624615	5.954208	8.853390	13.57058	9.250290
Skewness	1.664216	2.367956	2.648149	1.090103	-0.095769	15.21881	16.91272	36.98494	22.90042	16.24021	35.51542
Kurtosis	9.311994	6.607213	8.012692	2.188325	1.009172	400.0294	702.4228	2971.120	1940.042	516.7037	3847.799
Jarque-Bera	930505.1	647644.5	9711769.7	98900.80	73097.37	2.90E+09	8.96E+09	1.61E+11	6.86E+10	4.84E+09	2.70E+11
Probability	0.000000	0.000000	0.000000	0.000000	0.000000	0.000000	0.000000	0.000000	0.000000	0.000000	0.000000
Sum	1472522.	51759.00	44299.00	114341.0	229776.0	318211.0	434635.0	288831.0	705219.0	1053894.	592988.0
Sum Sq. Dev.	1592839.	45650.59	39824.51	84531.13	109392.9	6749633.	5761913.	15548584	34376532	80768057	37527843
Observations	438575	438575	438575	438575	438575	438575	438575	438575	438575	438575	438575

APPENDIX 2: THE RELATIONSHIP BETWEEN THE NUMBER OF PATENTS GRANTD AT THE BEGINNING OF THE SPELL AND THE SPELL LENGTH

Y = Spell length	Level regression	Log-log regression
Nb patent at the beginning	-0.109 (-2.829)	-0.102 (-0.788)
Intercept	1.084 (59.943)	12.549 (25.824)
N	337	337
R square	0.024	0.002
Adjusted R square	0.021	-0.01
F statistic	8.002	0.620

REFERENCES

Almeida, P., Kogut, B., (1997), "The Localization of Knowledge and the Mobility of Engineers in Regional Networks," Management Science, (1999), vol. 45, n. 7, p.p.905-917.

Archibugi, D., (1988), "In Search of a Useful Measure of Technological Innovation ," Technological Forecasting and Social Change, vol. 34, n° 3, pp. 253-277.

Archibugi, D., Michie, J., (1995), "The Globalization of Technology: A New Taxonomy," Cambridge Journal of Economics, vol. 19, n° 1, pp. 121-140.

Arundel, A., Kabla, I., (1999), "What percentage of innovation are patented? Empirical estimate for European firms," Research Policy, vol. 27, pp 127-141.

Florida, R., (2001), " The Economic Geography of Talent, , Annals of the Association of American Geographers, vol. 92, n° 4, pp. 743-755.

Gay, C., (2003), Économie de l'innovation technologique localisée. Un essai sur les organisations, communautés et individus apprenants, Thèse, soutenue le 18 décembre 2003. University of Lyon 2.

Gay ,C., Latham, B., Le Bas, C. (2005), "Collective Knowledge, Prolific Inventors and the Value of Inventions: An Empirical Study of French, German and British Owned U.S. Patents, 1975-1998." Under review at Economics of Innovation and New Technology for a special issue on the Governance of Technological Knowledge.

Griliches, Z., (1991), "The Search of RD Spillovers", NBER Working Paper, n° 3768.

Hall, B. H., Jaffe, A. B., Trajtenberg, M., (2001), "The NBER Patent Citations Data File: Lessons, Insights and Methodological Tools," in: JAFFE A. and M. Trajtenberg, Patents, Citations and Innovations. A window on the Knowledge Economy, MIT Press, pp. 403-459.

Trajtenberg, M., Patents, Citations and Innovations. A window on he knowledge Economy, MIT Press, pp. 403-459.

Howell, M., Boies, K., (2004), "Champions of technological innovation: The influence of contextual knowledge, role orientation, idea generation, and idea promotion on champion emergence," *The leadership Quarterly*, vol. 15, pp. 123-143.

Jaffe, A. B., Trajtenberg, M., (2002), *Patents, Citations and Innovations. A Window on the Knowledge Economy*, the MIT Press, Cambridge, Massachusetts, London.

Mumford, M. D., Scott, G. M., Gaddis, B., Strange, J. M., (2002), "Leading creative people: Orchestrating expertise and relationships,"*The Leadership Quarterly*, vol. 13 pp. 705-750.

Patel, P., (1995), "Localised Production of Technology for Global Markets," *Cambridge Journal of Economics*, vol. 19, n° 1, pp. 141-153.

Rothwell, R., (1992), "Successful industrial innovation: critical factors for the 1990s'," *R and D Management*, vol. 22, n° 3, pp.221-239.

Stolpe, M., (2001), "Mobility of Research Workers and Knowledge Diffusion as Evidenced in Patent Data. The Case of Liquid Crystal Display Technology," *Kiel Working Paper* n°1038.

Zucker, L. G., Darby, M. R., (1996), "Star Scientists and Institutional Transformation: Patterns of Invention and Innovation in the Formation of the Biotechnology Industry," *Proc. Natl. Acad. Sci. USA,* vol. 23, n° 93, pp. 12709-12716.

Zucker, L. G., Darby, M. R., (1998), "Capturing Technological Opportunity via Japan's Star Scientists: Evidence from Japanese Firms' Biotech Patents and Products," NBER Working Paper 6360.

Chapter 5

COMPARING INNOVATIVE PERSISTENCE ACROSS COUNTRIES: A COX-MODEL OF PATENTING IN THE UK AND FRANCE

Alexandre Cabagnols, *Clermont-Ferrand University and University of Lyon 2*

1. INTRODUCTION

The evolutionary framework defined by Nelson and Winter (1982) and Winter (1984) has initiated two related types of analysis concerning the determinants of the persistence in innovation: a first one at the industry level and a second one at the firm level. At the industry level firms' aggregate propensity to persist in innovation is supposed to be determined by the market structure and, more broadly, by the technological regime of the industry (Cohen and Levin (1989); Malerba and Orsenigo (1993), (1995); Van Dijk (2000); Breschi, Malerba and Orsenigo (2000)). At the firm level, persistence in innovation would be determined by two different feedbacks: a first feedback stemming from the economic success of the innovative activity in term of sales, profit and finally in term of market power; a second one stemming from the learning process associated to the technological activity. The first feedback results from the decision made to maintain the innovative effort and eventually to reinvestment a fraction of the profits generated by the innovation for the enlargement of production and research capacities. The second feedback stems from the accumulation of related technological knowledge and from the development of specific competencies of "learning to learn". Our paper focuses exclusively on that second technological feedback and studies its impact on the ability of French and UK firms[1] to persist in innovation.

Most of the analysis dealing with technological accumulation use R&D data that refer to the overall technological activity of the firm. Not only do they report a correlation between firms' performances (measured with productivity indicators, patent data or indicators of the share of new products in sales[2]) and current R&D expenses, but they also report a significant and positive impact of the cumulative level of R&D expenses on firms' performances (Hall and Mairesse (1995)). That last result is supposed to be the sign of a learning phenomenon reflecting positive feedbacks in the accumulation of technological knowledge. Our research addresses that question in a narrower way since it focuses exclusively on the impact of the past patenting activity of firms on their ability to keep on patenting in the US Patent and Trademarks Office during the 1969-84 period. Therefore, it exclusively intends to capture technological feedbacks emerging from and directed toward patented technological innovations in a specific patent office. Obviously, such a measure of the technological activity of firms only accounts for a limited fraction of the full technological output of innovative firms (Cohen, Nelson and Walsh (2000)).

Our datasets have already been explored in depth by Geroski, Van Reenen and Walters (1997) for the UK and Le Bas, Cabagnols and Gay (2001) for France. The motivation for a renewed interest for their work comes from the ambiguity of their conclusions (similar for French and UK firms): on the one hand they report a positive and significant impact of the initial number of patent on the probability of persistence in the patenting activity which may be considered as the sign of a learning process in innovation; on the other hand they indicate that the probability of persistence in the patenting activity decreases with time spent in that activity (the baseline hazard function is growing) which may be considered as contradictory with the idea of a learning process since in presence of a learning phenomenon, lengthening periods of success should lead to a lower probability of failure. The purpose of our work is to clarify that question whose implications are important for the economics of technological change.

Firstly, we suggest several hypotheses concerning the expected relationship between the level of technological accumulation of firms and their ability to keep on patenting. Secondly we describe the French and UK patent datasets available for that research. We suggest using a semi-parametric Cox model with time varying covariates rather than a parametric Weibull model. We detail the construction of the proxies used to measure the level of technological accumulation of firms. Thirdly, we present the results obtained from different estimations of a semi-parametric Cox model in which the level of technological accumulation of firms enters as a time varying covariate. A comparison between French and UK samples is performed in order to differentiate sample specific results from conclusions of a broader relevance.

2. TECHNOLOGICAL ACCUMULATION AND PERSISTENCE IN INNOVATION: AN EVOLUTIONARY VIEW

Firstly, we clarify the concept of persistence in order to introduce the specific terminology associated to that field of research. Secondly, we present two main premises inspired from the evolutionary literature on technological change that lead, thirdly, to suggest a number of hypotheses concerning the linkages between firms' ability to persist in innovation and their level of technological accumulation.

2.1 STATISTICAL APPROACHES TO PERSISTENCE IN INNOVATION

Persistence in innovation refers to firms' ability to keep on innovating. *"In the simplest statistical interpretation, the notion of innovative persistence can be defined as the conditional probability that innovators at time t will innovate at time t+1. More precisely one can think of persistence as the degree of serial correlation in innovative activities"* Malerba, Orsenigo and Peretto (1997), p.804 Lets consider a variable x standing for the innovative behavior with two values j:{0, 1} where x=1 stands for innovation; x=0 stands for non-innovation. In a conditional probability language expressed in discrete time, persistence between t and t+n in state x=j is the probability for a firm i to remain in the same state j as expressed in equation 1.

$$persistence_{i,t\ldots t+n} = prob\left(x_{i\,t+n} = j \middle/ x_{it} = j \right) \quad \text{(Eq.1)}$$

This equation refers to a markov-chain perspective that can also be expressed in term of auto-regressive process (AR). It is of interest when, for example, we compare the level of persistence in innovation and in non-innovation or when we study transition probabilities between different types of innovative behaviors (Cefis and Orsenigo (2001), Cabagnols (2000)). Persistence can be studied in a second way that explicitly aims at shedding light on its evolution as the time spent in innovation gets longer (Geroski, Van Reenen, Walters (1997), Le Bas, Cabagnols and Gay (2001)). From that point of view, persistence is investigated using a specific conditional probability called "hazard rate" (*h*). *h* is supposed to vary with the time (*t*) that a firm *i* in country *w* has already spent in innovation:

$$h_{iw}(t) = \lim_{dt \to 0} \frac{prob\left(t \le T_{iw} < t+dt \middle/ T \ge t \right)}{dt} \quad \text{(Eq. 2)}$$

where T is the time at which the exit from innovation occurs (T is to whole duration of the innovative spell before failure[3]). Consequently h(t) is the probability for the failure to occur between time t and t+dt conditional on the fact that the failure has not been observed before t (i.e. T≥t).

The level of persistence is thus the complementary to 1 of h(t): the higher h(t), the lower persistence, and conversely. Our work relates to this second kind of approach particularly suited to identify the shape of the persistence process and to measure the impact of a number of explanatory variables.

2.2   EVOLUTIONARY PREMISES

The evolutionary tradition considers that the process of technological learning is characterized by its cumulativeness along path-dependent trajectories. In that sense firms' innovative behavior can be considered as self-fuelled and partly endogenous[4]. Following the evolutionary literature we try to suggest a simple approach to identify the likely impact of that internal process of technological accumulation on firms' ability to persist in innovation.

We suggest three main premises borrowed to the evolutionary literature:

Firstly, as it has been pointed out by Nelson and Winter (1982), Winter (1984), Dosi (1988) and extensively studied by Cohen and Levinthal (1990) the learning activity leading to innovation is partly self-fuelled through a process of technological accumulation. It means that "today innovations and innovative activities form the base and the building blocks of tomorrow innovations and that today innovative firms are more likely to innovate in the future" (Malerba and Orsenigo (1993), p.48). That phenomenon is mainly explained by the existence of technological complementarities/proximities between successive generations of technologies. In such a context, all other factors remaining equal, new innovators would face higher level of hazard due to their lack of previous technological experience. With time, persistent innovators would however progressively accumulate knowledge and therefore reinforce their ability to persist in innovation. The relevance of that analysis is confirmed by numerous empirical studies that report: 1/ a strong persistence of technological leadership between firms provided that technological complementarities and appropriability conditions are strong enough (Gruber, 1992, 1994); 2/ a significant and positive serial correlation between current and future technological performances of firms measured with patent data (Crépon and Duguet (1997)) and survey data (Cabagnols (2000); Martinez-Ross and Labeaga (2002); Duguet, Monjon (2002)); 3/ the existence of a quite stable core of innovators in most of the industrial sectors that also account for a disproportionate share of the innovative activity (Malerba and Orsenigo (1999)).

Secondly, with the rise of the level of technological accumulation at the firm level, major opportunities may progressively be exhausted and some lock-in/lock-out effect may take place resulting in a reduction of firms' ability to grasp new technological opportunities emerging in different technological fields ((Cohen and Levinthal (1990), March (1991)). Therefore, technological accumulation along a same trajectory may exhibit decreasing returns in term of economic gains, technological advance and innovative persistence. As pointed out by March (1991) such a process can however be avoided if firms manage efficiently the trade-off between technological exploitation and technological exploration. The balance between technological diversity and technological specialization may therefore be a fundamental explanatory variable of firms' ability to keep on innovating when their level of technological accumulation grows larger.

Thirdly, as time spent in innovation lengthens a phenomenon of "learning to learn" may take place. It has been widely mentioned in the evolutionary literature. That "learning to learn" process designates the fact that with time firms learn how to use more efficiently their technological resources for the development of subsequent innovations. Such a phenomenon may result in a time varying impact of the level of technological accumulation of firms on their probability of persistence in innovation. In particular, the learning to learn phenomenon would explain a progressive increase of the impact of the level of technological accumulation on persistence that reduces the risk of lock-in and opens-up new opportunities.

All other factors remaining equal[5], the combination of these three fundamental hypotheses brings to the conclusion that persistence in innovation would follow a widely observed shape: initially a fast increase due to a process of technological accumulation and "learning to learn"; progressively the rise of an inhibition factor (lock-in/lock-out) leading the process to a saturation threshold[6] if firms do not diversify their technological activity and if their level of ability to learn has not sufficiently increased to circumvent the risk of lock-in.

2.3   HYPOTHESES

Using the theoretical approach that has just been described, we present a number of hypotheses. The first set of hypotheses deals with the impact of the level of technological accumulation on firms' persistence in innovation. It will be explicitly tested latter. The second set of hypothesis deals with the expected shape of the hazard function if the level of technological accumulation and the "learning to learn" phenomenon are the main driving forces of the persistence process.

<u>Impact of the level of technological accumulation (K)</u>

The evolutionary theory states that the technological learning is cumulative. It means that firms' ability to develop future technological innovations (in t+dt) relies heavily on their current level of technological knowledge (labeled K_t). Consequently, we expect a positive impact of K_{ti} on the probability of persistence in innovation (i.e. a negative impact on the hazard rate). As a result our first hypothesis is: $H_1: \frac{\partial h_t}{\partial K_t} < 0 \ \forall t$ which means that a larger level of technological accumulation leads to a lower hazard rate (probability to exit innovation).

However, as indicated by several authors, in order to benefit from that accumulated knowledge, firms have to keep on searching in related/complementary technological fields. The problem is that such a strategy often results in lock-in effect, incremental innovations (as opposed to radical innovations) and in a progressive exhaustion of the technological opportunities in the field of technological activity of the firm. Consequently we expect that the positive impact of the technological accumulation is not linear but positive at a decreasing rate, which measures the intensity of the saturation process that has just been mentioned. A second hypothesis can thus be suggested: $H_2: \frac{\partial^2 h_t}{\partial K_t^2} > 0 \ \forall t$. Combined with H_1, it means that a higher level of technological accumulation increases at a decreasing rate the probability of persistence in innovation.

Turning now to the "learning to learn" process, its impact is revealed by a more efficient use by firms of their stock of technological knowledge. Consequently, we suggest a third hypothesis: $H_3: \frac{\partial}{\partial t}\left(\frac{\partial h_t}{\partial K_t}\right) < 0 \ \forall t$.

Combined with H_1, it means that with the lengthening of the innovative spell, the benefit (in term of persistence) resulting from a same increase in the level of technological accumulation is reinforced since firms learn how to use it more efficiently for innovation.

3. ECONOMETRIC MODELLING OF FIRMS' PERSISTENCE IN INNOVATION

Whereas our work uses the same dataset as Geroski et al. (1997) and Le Bas et al. (2001), we do not carry out the same econometric treatment. We have made three main changes: firstly different rules are applied for the construction of the datasets; secondly, rather than modeling the hazard rate with a parametric Weibull specification we suggest using a semi-parametric Cox specification[7]; thirdly, rather than focusing on the shape of the hazard function in order to study a possible "learning phenomenon" we try to catch-it directly trough the introduction of explanatory variables. Justifications for that strategy are detailed below.

3.1 BUILDING OF THE FRENCH AND UK DATABASES

POPULATION

We use two datasets reporting the number of patents granted by the US Patent and Trademark Office (US-PTO) to French and UK firms. In both cases public research institutions, public firms and individual patentees were dropped; the focus is thus exclusively on private firms. For UK we use the same dataset as Geroski, Van Reenen and Walters (1997). It is a balanced panel 3304 firms selected from the population of over 7500 UK firms that were granted patents by the United States Patents and Trademarks Office at some time in the 1969-1988 period. Only private and quoted firms were included in that dataset. For France we use a dataset collected by the SPRU that exhaustively reports French entities that were granted patents by the US-PTO at some time during the 1969-84 period. From that database we extracted private firms either quoted or non-quoted.

The inclusion of non-quoted firms in the French dataset makes certainly the average firm size of that dataset lower than in that of the UK one. Such differences in size may result in specific processes of technological learning: smaller firms may be more sensitive to technological accumulation than larger firms and may develop a quicker "learning by learning" (Duguet and Monjon 2002). Consequently the main interest of that France/UK comparison is not to perform a precise international evaluation but to make sure that our conclusions concerning a possible learning process in the accumulation of technological knowledge related to the patenting activity are not strictly country and sample specific.

INNOVATION AND PATENTING BEHAVIORS

As indicated by several empirical analyses, technological innovations are not always patented (Levin, Klevorick, Nelson et Winter (1987), Griliches (1990); Harabi (1995), Duguet and Kabla (1998); Cohen et al. (2000)). The propensity to patent in a given patent office is mainly influenced by firm specific characteristics (in particular its size and the geographic location of its market), the type of technological innovation (product or process; incremental or radical), industry specific conditions of innovation (in particular the appropriability conditions). Consequently patent statistics do not fully report the innovative activity of firms; in many cases innovations will not be patented or will be patented in a different patent office than the one under interest (in our case the US-PTO). Conversely, patent statistics can lead to an overestimation of the innovative activity since patents have also a strategic function and often never end-up with a real economic application of the corresponding underlying technological knowledge (Cohen et al. (2000)). Consequently the scope of our conclusions is limited to the subset of French and UK firms for which the US patent strategy is relevant and actually used. Although we know that the relationship between patenting and innovative behaviors is not perfect[8] we will suppose that persistence in patenting behaviors can be at least related to a persistent "manifested capability of innovation".

SPELL LENGTH AND CENSORING

Definition of a spell

These two datasets report yearly the number of patents granted to French and UK firms in the US-PTO. Despite they do not apply that procedure to their dataset, Geroski, Van Reenen and Walters (1997)[9] suggest considering that a patenting spell ends only if a firm stops patenting 2 years or more (the "lag" between two successive years of patenting activity should be larger than or equal to 2 years; lag≥2). After a careful analysis of the sensitivity of the hazard function to the value of that criterion we have decided not to use that 'fill-in' procedure since it strongly and qualitatively modifies the shape of the hazard function in a manner that is proportional to the value of the lag that is used.

With the "more than one year" criterion (lag≥2) we observe a jump of the hazard rate in year 3 whereas with the "more than two years" criterion (lag≥3) we observe a leap of the hazard function in year 4. Generally speaking the criterion of interruption induces an artificial surge of the hazard rate in time t=lag+1. In order to avoid any artifactual phenomenon we have decided to consider that a spell ends after one year without patenting activity (lag≥1).

Graph 1: Kaplan-Meier estimate of the French and UK hazard functions for different values of lag

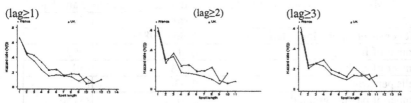

(lag≥1)　　　　　　　　　　(lag≥2)　　　　　　　　(lag≥3)

Censoring procedure

Given the nature of our problematic, entry conditions into the patenting activity have to be clearly defined in order to avoid major measurement errors concerning the stock of technological knowledge accumulated by firms during the spell (K_t). An important source of error in the measurement of K_t comes from our incapacity to identify the starting date of a spell. Such a problem arises when firms begin patenting right from the first year of coverage of the dataset (1969). Such observations are considered as left censored. Doing as if 1969 was the first year of patenting activity leads to an obvious underestimation of the stock of technological capital of the firms that were already patentees before 1969[10]. Moreover, as indicated earlier, a one year period without patent (in that case 1969) may be purely fortuitous and not related to a true interruption of the underlying technological and patenting activity. Consequently, we have decided to suppress form our database any spell starting in 1969 or 1970. These two years of control make us sure that firms under investigation are really engaged in a new spell of patent since they were not granted patents during at least two consecutive years before their first entry in the database. As a result, the stock of technological knowledge accumulated by firms through each of the spells in which they are engaged can be correctly measured. That restriction to spells starting in 1971 and after results in the elimination from the database of many important patentees that account for a larger number of patents than *"new patentees"* starting a spell after 1970. Such a result indicates that measurement errors concerning K_t for left censored observations were certainly very important and were likely to induce estimation biases[11].

In order to avoid any additional distortion in the comparison of UK and French firms due to different periods of investigation we use the year 1984 as the last date of observation both for France and UK[12]. Consequently our period of investigation is set to [1971-1984]; it remains in the dataset 2951 firms for France and 2481 for UK. We notice that the loss of observations due to that treatment is more important in the UK dataset than in the French one. It reveals that the problem of left censoring and measurement errors of K_t were certainly more important in the case of UK.

Table 1: Main characteristics of the dataset used to compare France and UK

(lag≥1)	France	UK
Left censored spells included	[1969-1984]	
Number of firms	3347	2962
Number of spells	4468	4502
Spells per firm	1.33	1.52
Number of patents granted	22044	28156
Patents per firm	6.59	9.51
Left censored spells excluded	[1971-1984]	
Number of firms [% loss *]	2951 [12%]	2481 [16%]
Number of spells [% loss]	3832 [14%]	3584 [20%]
Spells per firm [% loss]	1.3 [2%]	1.44 [5%]
Number of patents granted [% loss]	10662 [52%]	9490 [66%]
Patents per firm [% loss]	3.61 [45%]	3.83 [60%]

*% loss refers to the difference between the results obtained with and   without left censored observations.

3.2   SHAPE OF THE HAZARD FUNCTION

DESCRIPTIVE ANALYSIS OF THE HAZARD FUNCTION

The first step of a survival time analysis is the descriptive presentation of the hazard function using the Kaplan-Meyer technique. The Maximum Likelihood estimate provided by that technique for the hazard rate is $\hat{h}j = d_j \big/ n_j$ and it is $\hat{S}(t) = \prod_{j/t_j < t}\left(\dfrac{n_j - d_j}{n_j}\right)$ for the survival function; where d_j stands for the number of failures at time t_j; n_j is the total number of subjects at risk just prior t_j; t_k is the observation time when failures occur (see Kalbfleisch and Prentice (1980), p.11-12). Estimates of $\hat{S}(t)$ and $\hat{h}j$ for France and UK are reported below.

Table 2: Kaplan-Meier estimate of the survival functions in France and UK (spell interruption after one year exactly or more without patent; lag≥1)

	Total Remaining (nj)	Net Fail (dj)	Lost due to censoring (mj)	Survivor Function S(t)	Standard Error	95% Conf. Int. for the survivor function		Hazard function h(t)
Current spell length tj	France - 1971-84 (applying our own treatment to the dataset)							
1	3832	2734	313	0.2865	0.0073	0.2723	0.3009	0.713
2	785	364	60	0.1537	0.0064	0.1413	0.1665	0.464
3	361	131	26	0.0979	0.0056	0.0872	0.1093	0.363
4	204	52	16	0.0729	0.0052	0.0633	0.0835	0.255
5	136	22	14	0.0611	0.0049	0.0520	0.0712	0.162
6	100	19	5	0.0495	0.0046	0.0410	0.0592	0.190
7	76	13	12	0.0411	0.0044	0.0330	0.0503	0.171
8	51	7	5	0.0354	0.0043	0.0277	0.0445	0.137
9	39	5	3	0.0309	0.0042	0.0234	0.0399	0.128
10	31	4	9	0.0269	0.0041	0.0197	0.0358	0.129
11	18	1	4	0.0254	0.0041	0.0182	0.0345	0.056
12	13	0	7	0.0254	0.0041	0.0182	0.0345	~
13	6	0	3	0.0254	0.0041	0.0182	0.0345	~
14	3	0	3	0.0254	0.0041	0.0182	0.0345	~
Current spell length tj	UK - 1971-84 (applying our own treatment to the dataset)							
1	3584	2489	246	0.3055	0.0077	0.2905	0.3207	0.694
2	849	412	63	0.1573	0.0066	0.1446	0.1704	0.485
3	374	171	18	0.0854	0.0054	0.0752	0.0963	0.457
4	185	65	11	0.0554	0.0046	0.0468	0.0649	0.351
5	109	27	10	0.0417	0.0042	0.0340	0.0503	0.248
6	72	18	2	0.0312	0.0038	0.0245	0.0393	0.250
7	52	8	5	0.0264	0.0036	0.0201	0.0341	0.154
8	39	8	2	0.0210	0.0033	0.0152	0.0282	0.205
9	29	5	2	0.0174	0.0031	0.0121	0.0243	0.172
10	22	2	2	0.0158	0.0030	0.0107	0.0226	0.091
11	18	4	4	0.0123	0.0028	0.0077	0.0188	0.222
12	10	1	1	0.0111	0.0028	0.0066	0.0177	0.100
13	8	1	3	0.0097	0.0028	0.0053	0.0164	0.125
14	4	0	4	0.0097	0.0028	0.0053	0.0164	~
Current spell length tj	UK - 1969-88 (applying the same treatment to the data as Geroski et al (1997))							

	Total Remaining (nj)	Net Fail (dj)	Lost due to censoring (mj)	Survivor Function S(t)	Standard Error	95% Conf. Int. for the survivor function		Hazard function h(t)
1	5178	3544	177	0.3156	0.0065	0.3029	0.3283	0.684
2	1457	690	40	0.1661	0.0053	0.1558	0.1767	0.474
3	727	275	24	0.1033	0.0045	0.0947	0.1123	0.378
4	428	123	12	0.0736	0.0039	0.0662	0.0815	0.287
5	293	67	21	0.0568	0.0035	0.0502	0.0639	0.229
6	205	32	7	0.0479	0.0033	0.0417	0.0547	0.156
7	166	27	7	0.0401	0.0031	0.0344	0.0465	0.163
8	132	11	1	0.0368	0.003	0.0312	0.043	0.083
9	120	12	4	0.0331	0.0029	0.0278	0.0391	0.100
10	104	10	1	0.0299	0.0028	0.0248	0.0357	0.096
11	93	17	3	0.0244	0.0026	0.0198	0.0298	0.183
12	73	3	2	0.0234	0.0025	0.0189	0.0288	0.041
13	68	5	1	0.0217	0.0024	0.0173	0.0269	0.074
14	62	9	1	0.0186	0.0023	0.0144	0.0235	0.145
15	52	3	1	0.0175	0.0023	0.0135	0.0223	0.058
16	48	10	1	0.0138	0.0021	0.0102	0.0184	0.208
17	37	7	1	0.0112	0.0019	0.008	0.0154	0.189
18	29	3	3	0.0101	0.0018	0.007	0.0141	0.103
19	23	1	0	0.0096	0.0018	0.0066	0.0137	0.043
20	22	0	22	0.0096	0.0018	0.0066	0.0137	0.000

Graph 2: Kaplan-Meier estimate of the
Survivor function (lag≥1)

Graph 3: Kaplan-Meier estimate of the
Hazard function (lag≥1)

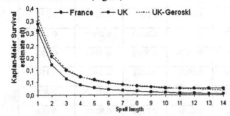

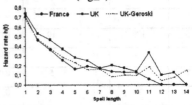

Tests of equality of the French and UK survivor functions (log-rank test
and Wilcoxon test) do not report significant differences between French and
UK survivor functions[13]: It indicates that French and UK populations of
patenting firms follow a similar process of persistence in innovation over the
1971-84 period. Moreover, we notice that our treatment of the data does not
lead to important departures from the results obtained with the treatment of
Geroski et al. (1997). The main difference induced by our treatment is an
increase of the hazard rate all along the spell which results in lower survivor

probabilities. That change was expected since the removal of left censored spells from the dataset leads to the elimination of many very large innovative firms that were certainly long lasting innovators and had accumulated a very large (but not precisely determined) "stock of technological knowledge" which is supposed to enhance the probability of persistence in innovation.

The visual evaluation of the Kaplan-Meier estimation indicates that the hazard function is strongly decreasing. However, it does not prove that a learning process resulting from a process of technological accumulation is the driving force of that phenomenon since:
- Firstly, in addition to learning, other uncontrolled explanatory variables may be at work that explain the decreasing trend of the hazard rate.
- Secondly, if not controlled, the grouping of heterogeneous sub-populations of individuals in a same population leads mechanically to a decreasing hazard function. That problem is known as a question of "unobserved heterogeneity"[14]. Since discrimination between potentially heterogeneous sub-populations is not always easy (due to the very large number of possible sources of heterogeneity between individual firms), we can't assert that the true hazard rate is not simply constant or following another non-linear form. Consequently, any conclusion based on the descriptive analysis of the shape of the hazard function may be extremely misleading.

PARAMETRIC ANALYSIS OF THE HAZARD FUNCTION

Usually the introduction of explanatory variables in a survival analysis is made through the estimation of parametric models that rely on an a priori specification of the functional form of the hazard function. Econometricians prefer using such parametric models since they make possible to perform hypothesis tests on the shape and lead to a "gain in efficiency" (Geroski et al. (1997), p.36). However they raise a problem of selection in order to determine which is the most suited functional form. If the wrong specification is chosen estimations may lead to misleading conclusions concerning both the shape of the hazard function and/or the coefficients associated to the explanatory variables (Kalbfleisch and Prentice (1980), p.67). In such situations the Kaplan-Meier estimate of the hazard functions is often used as a benchmark for the evaluation of the hazard function since its states no a priori hypothesis concerning its functional form.

In a first approach, in order to assess the relevance of a parametric specification we have estimated several parametric models without covariates at the exception of one constant (Poisson, Weibull, Loglogistic and lognormal[15])[16]. Afterwards we have compared the estimated hazard functions produced by these models to the hazard function obtained with a Kaplan-Meier approach. Except for the Poisson model of which hazard function is constant ($h=\lambda$) all of them can handle with decreasing hazard functions so that it is difficult to determine a priori which is the best one:

The Weibull model with a hazard function of the form $h=\lambda p\left(\lambda t\right)^{p-1}$ is the most widely used[17]. It is particularly suited to model situations of constant hazard (p=1), monotone increasing hazard (p>1) and monotone decreasing hazard (p<1). Given former results obtained with the Kaplan-Meier approach and our theoretical hypothesis we expect p<1 (i.e. a decreasing hazard function) when none covariate is entered into the model.

The lognormal model where

$$h_t = \frac{\dfrac{1}{\sqrt{2\pi}\,t}\,p\,e^{\left(\frac{-p^2\left(\log(\lambda t)^2\right)}{\acute{e}}\right)}}{1-\Theta\left(p\log(\lambda t)\right)} \text{ with } \Theta(w)=\int_{-\infty}^{w}\frac{e^{-w^2/2}}{\sqrt{2\pi}}du \text{ is suited to handle}$$

function with increasing then decreasing hazard rates. The hazard is 0 for t=0; it increases up to a maximum after which it tends to 0 as t goes to $+\infty$.

The log-logistic model looks like the lognormal one but in a simpler way:

$$h_t=\frac{\lambda p\left(\lambda t\right)^{p-1}}{1+\left(\lambda t\right)^{p}}.$$ Depending on the value of p the hazard follows different

trajectories: if p<1 monotone decreasing; if p>1 growing from 0 to a maximum and afterwards decreasing to 0 Table 3 reports results from the estimation of these different specifications for France and the UK separately.

Table 3: Separate estimations of the spell length in France and UK

	Model			
	Exponential	**Weibull**	**Lognormal**	**Log-logistic**
	France			
Constant $(\lambda=e^{-constant})$	**.5229**	**.5888**	**.2675**	**.14353**
Standard deviation	.0176	.0156	.0094	.0082
Ancillary (p)		**1.3740**	**.511**	**.2312**
Standard deviation		.0274	.0122	.0073
Log-likelihood	-4504.82	-4174.24	-2900.00	-2443.05
AIC	9011.66	8352.48	5804.02	4890.12
Shape of the estimated hazard function	*Constant*	*Increasing*	*Decreasing*	*Decreasing*

	UK			
Constant $(\lambda=e^{-constant})$	**.5114**	**.5879**	**.2817**	**.1684**
Standard deviation	.0164	.01494	.0092	.0089
Ancillary		**1.4366**	**.5023**	**1/.2428**
Standard deviation		.03129	.0110	.0066
Log-likelihood	**-4151.56**	**-3751.92**	**-2628.28**	**-2329.55**
AIC	8305.14	7507.84	5260.58	4663.12
Shape of the estimated hazard function	*Constant*	*Increasing*	*Decreasing*	*Decreasing*

Akaike Information Criterion (AIC) = -2log(log likelihood) + 2(number of ancillary parameters+1)

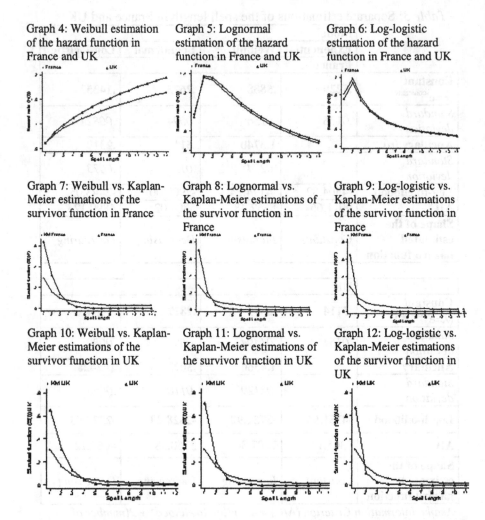

Graph 4: Weibull estimation of the hazard function in France and UK

Graph 5: Lognormal estimation of the hazard function in France and UK

Graph 6: Log-logistic estimation of the hazard function in France and UK

Graph 7: Weibull vs. Kaplan-Meier estimations of the survivor function in France

Graph 8: Lognormal vs. Kaplan-Meier estimations of the survivor function in France

Graph 9: Log-logistic vs. Kaplan-Meier estimations of the survivor function in France

Graph 10: Weibull vs. Kaplan-Meier estimations of the survivor function in UK

Graph 11: Lognormal vs. Kaplan-Meier estimations of the survivor function in UK

Graph 12: Log-logistic vs. Kaplan-Meier estimations of the survivor function in UK

The AIC criterion indicates that the log-logistic model is the best one both for France and UK. However, if we look at expected *vs.* observed[18] survival functions, we immediately notice the poor adjustment of these parametric models to the data (see Graph 7 to Graph 12 above). We observe an important underestimation of the hazard rate during the first years whereas afterwards it is systematically overestimated. A specific analysis of the residuals confirms this result. The Weibull model seems particularly misleading since it reports a continuously growing hazard rate[19]. Lognormal and log-logistic models perform better in term of AIC but contrary to the initially decreasing Kaplan-Meier estimate of the hazard function, they report firstly a growing hazard rate between the first and the second years that afterwards decreases.

We have performed the same type of analysis after introduction of categorical explanatory variables. In most cases the estimation the ancillary parameter that account for the shape of the hazard function was not in line with the results obtained via a stratified estimation of the Kaplan-Meyer descriptive hazard function. Our conclusion is that the functional form of the hazard rate of the patenting process of our dataset is not easily grasped with a usual specification.

3.3 RELEVANCE OF THE COX MODEL

In order to rule out the problem of the functional specification of the hazard function we have decided to use a Cox proportional hazard model. The interest of this model is that it does not specify a priori a functional form for the baseline hazard function and consequently avoids any misleading conclusions concerning the shape of the hazard function when it exists a doubt concerning the right functional specification[20]. The Cox model considers that the baseline hazard function is a priori unknown. Rather than using raw duration data, the model uses ranks of the duration in order to estimate firstly the impact of the explanatory variables. Descriptive baseline hazard and survivor functions are calculated in a second step in which the estimated coefficients associated to the covariates are used as parameters[21]. These functions represent the survival rate and the level of hazard that firms face when all the covariates entered into the model are set to 0.

In its basic version the semi-parametric Cox model rests upon the hypothesis of proportional hazard[22] and therefore handles only time-independent explanatory variables[23]. It is often written as follows:

$$h(t,X) = h_0(t)e^{\sum \beta_i X_i}$$ (Eq. 3) where:

$h_0(t)$ is the baseline hazard rate defined for each observation time t. It is not a priori specified with a function.

x_i's are time-independent covariates (fixed once for all at the beginning of the spell).

That model can be stratified in order to take account of different baseline hazard functions for different sub-populations. In that case the model can be

rewritten as it follows: $h(t,X) = h_{0w}(t)e^{\sum \beta_i X_i}$ (Eq. 4) where w figures out a stratification variable (in our case we will stratify by country).

A further extension of that model consists in the introduction of time dependent covariates[24]. In that case the model is reformulated as it follows:

$$h(t,X(t)) = h_{0w}(t)e^{\sum_{i=1}^{k} \beta_i X_i + \sum_{j=1}^{l} \beta_j X_j(t)}$$ (Eq. 5) where $x_j(t)$'s are time-dependent explanatory variables.

From the point of view of the economics of technological change the synthetic parameter p previously provided by the Weibull model is no longer estimated. Rather, in a Cox specification, the impact of a possible process of technological learning has to be explicitly measured with a covariate. That covariate has to be considered as time-varying due to the combined action of the obsolescence of former technological knowledge and the accumulation of new knowledge gained trough recent technological activities. Consequently the most relevant model for our purpose is a Cox model with time varying covariates (Eq.5 above).

3.4   MEASUREMENT ISSUE AND MODEL SPECIFICATION

The implementation of the Cox model described above faces two important measurement issues: firstly we have to build a proxy measure for the level of technological knowledge accumulated by firms; secondly we must indicate how that variable can be introduced into the model.

MEASUREMENT OF THE LEVEL OF TECHNOLOGICAL ACCUMULATION

Firms' level of "technological accumulation" is an abstract concept that is not directly observable. Usually that technological capital is measured using a perpetual inventory technique[25] that takes account of an obsolescence parameter. Since our study deals with patent data, we will approximate the level of technological accumulation of a firm i in country w at time u with:

$$K_{u_{iw}} = \sum_{j=0}^{j=u} k_{uiw}.(1 - \delta_w)^{u-j} \quad \text{(Eq. 6) where:}$$

k_{jiw} is the number of patents granted to the firm i at time j in country w[26].

δ_w is the obsolescence rate of the technological knowledge in term of relevance for further innovations in country w[27].

$j=0$ is the first year of patenting activity of the firm when we consider its whole history.

Since we define a spell as a "non-stop period of patenting activity" we can distinguish two components in $K_{u_{iw}}$:

The technological accumulation made prior the beginning of the current spell (i.e. for j<(start-lag)).

The technological accumulation made during the current innovative spell and until now (i.e. for start<j≤u ⟺ 0<s<t where t=u-start is the time spent in the current spell).

Therefore, Eq. 6 can be rewritten:

$$K_{u_{iw}} = \sum_{j=0}^{j=start-lag-1} k_{jiw} \cdot (1-\delta_w)^{u-j} + \underbrace{\sum_{j=start}^{j=u} k_{jiw} \cdot (1-\delta_w)^{u-j}}_{K_{t_{iw}} = \sum_{s=0}^{s=u-start=t} k_{siw} \cdot (1-\delta_w)^s} \quad \text{(Eq. 7)}$$

where *lag* is the number of years without patents between the previous and the current spell.

Each part of that equation will be introduced as a specific explanatory variable:

The left hand side of Eq.7 refers to the past (or "pre-spell") technological experience of the firm. It is labeled EXPE. We suppose that for each firm EXPE is a fixed technological asset at the beginning of the spell but that its impact on the probability of persistence is likely to change depending on the lag between the last period of patenting activity and the starting date of the current spell. That variable aims at capturing inter-spell transfers of technological knowledge. We will not directly measure the value of EXPE as a sum of patents. Instead, we will capture its impact using three dummy variables EXP_1, EXP_2, EXP_3:

$EXP_1=1$ if the firm experienced a previous patenting spell less than 4 years ago (lag≤3).
$EXP_2=1$ if the firm experienced a previous patenting spell between 4 and 6 years ago (4≤lag≤6);
$EXP_3=1$ if the firm experienced a previous patenting spell more than 6 years ago (lag>6).
$EXP_1=EXP_2=EXP_3=0$ if no previous patenting activity is reported in the database[28].

We expect a positive impact of each of these dummies on the probability of persistence (i.e. negative on the hazard rate). The introduction of 3 dummies rather than of only one enables to check the hypothesis of a positive obsolescence rate of the technological knowledge (δ): we think that the impact of EXP_1 will be larger than the impact of EXP_2 and EXP_3.

The right hand side of Eq.7 K_t represents the level of technological accumulation made throughout the current spell after t years of non-stopped patenting activity in the US-PTO. That variable measures the "intra-spell" technological accumulation. It is labeled $ACCU_t$. Its value changes each year depending on the level of patenting activity of the firm during the spell and depending on the obsolescence rate. Since left censored observations have been removed from the dataset, no measurement error is made on that variable. $ACCU_t$ is a time varying variable (varying with t) that can be introduced differently in the model depending on the hypothesis under investigation:

$$H_1 : \frac{\partial h_t}{\partial K_t} < 0 \ \forall t \quad \text{It can be tested by introducing directly } ACCU_t \text{ as an}$$

explanatory variable of the hazard rate. H_1 will not be rejected if the estimated coefficient associated to $ACCU_t$ is negative and significant.

$$H_2: \frac{\partial^2 h_t}{\partial K_t^2} > 0 \ \forall t \quad \text{It can be tested by introducing } ACCU_t \text{ squared}$$

($ACCU_t^2$). H_2 will not be rejected if the estimated coefficient associated to $ACCU_t^2$ is positive and significant.

$$H_3: \frac{\partial}{\partial t}\left(\frac{\partial h_t}{\partial K_t}\right) < 0 \ \forall t. \text{ It can be tested with two main methods:}$$

We can introduce an interaction effect between the length of the current spell (*t*) and $ACCU_t$ that we label $ACCU_t \times t$. H_3 will not be rejected if the estimated coefficient associated to $ACCU_t \times t$ is negative and significant.

It is possible to build several dummy variables labeled $ACCU_{t1}$, $ACCU_{t2}$ and $ACCU_{t3}$ where:

$ACCU_{t1} = ACCU_t.g_1(t)$ where $g_1(t)=1$ if $t \leq 3$; $g_1(t)=0$ otherwise;
$ACCU_{t2} = ACCU_t.g_2(t)$ where $g_2(t)=1$ if ($t \geq 4$ and $t \leq 6$); $g_2(t)=0$ otherwise;
$ACCU_{t3} = ACCU_t.g_3(t)$ where $g_3(t)=1$ if $t > 6$; $g_3(t)=0$ otherwise.

H_3 will not be rejected if the estimated coefficient associated to $ACCU_{t1}$ is significantly lower than $ACCU_{t2}$ and than $ACCU_{t3}$. Simultaneously we can also test H_1 that will not be rejected if the estimated coefficients associated to $ACCU_{t1}$, $ACCU_{t2}$ and $ACCU_{t3}$ are all negative and significant.

Two important assumptions are made in that analysis: 1/at each spell a new period of technological accumulation starts; 2/the impact of the pre-spell technological activity of the firm on the hazard rate can be correctly captured by a fixed-effect. Said differently, we assume that a new cycle of technological accumulation starts at each spell whose shape is believed not to be qualitatively modified by the pre-spell technological experience of the firm. As indicated in appendix 4 that assumption can be considered as realistic from a statistical point of view.

Therefore, the scope of our econometric analysis will be limited to the shape of the hazard function observed during each spell of patenting activity and not over the whole potential period of patenting activity of the firm.

INTRODUCTION OF THE COVARIATES IN THE COX MODEL

We have estimated two models:

Model 1: $h(t) = h_{0w}(t) \cdot e^{\psi_{1w}EXPE_{1w} + \psi_{2w}EXPE_{2w} + \psi_{3w}EXPE_{3w} + \alpha_w ACCU_{tw} + \beta_w ACCU_{tw}^2 + \gamma_w ACCU_{tw} \times t}$

Model 2: $h(t) = h_{0w}(t) \cdot e^{\psi_{1w}EXPE_{1w} + \psi_{2w}EXPE_{2w} + \psi_{3w}EXPE_{3w} + \alpha_w ACCU_{t1w} + \beta_w ACCU_{t2w} + \gamma_w ACCU_{t3w}}$

Subscript w stands for the country: different baseline hazard functions ($h_{0w}(t)$) and different coefficients have been estimated for each country. The estimation procedure however has been made simultaneously for both countries, which makes possible to perform comparison tests of coefficients between countries (coefficients estimates and the baseline hazard function are not modified by that procedure).

Each model treats in the same way the past technological experience of the firm using the three previously described dummy variables (EXPE$_1$, EXPE$_2$, EXPE$_3$). These variables are fixed once for all at the beginning of each spell.

Differences between model 1 and 2 come from the way we measure K_t the "level of technological accumulation made throughout the spell". The first model introduces directly the "level of technological accumulation" (ACCU) of French and UK firms as a continuous time varying covariate (as indicated by the subscript t): ACCU$_t$, ACCU$_t^2$, ACCU$_t \times t$. In the second model the level of technological accumulation is split in three variables in order to focus exclusively on a possible time varying impact of the level of technological accumulation during the spell (ACCU$_{t1}$, ACCU$_{t2}$, ACCU$_{t3}$).

4. IMPACT OF THE LEVEL OF TECHNOLOGICAL ACCUMULATION ON PERSISTENCE IN INNOVATION: RESULTS FROM A COX MODEL

That third section reports results obtained from the estimation of the two Cox models described above. We firstly define the value of the obsolescence rate δ before performing a detailed analysis of the estimated coefficients associated to the covariates.

As indicated earlier our two models take account of country specificities denoted by the subscript w (1 for France, 2 for UK). The first country specificity lies in the stratification by country: two separate baseline hazard function ($h_{0w}(t)$) are estimated. The second country specificity is captured by the estimation of different coefficients for each explanatory variable in both countries. In that analysis time is measured in a discrete way. It results in many ties observations which have been handled with an exact method. A problem of correlation between observations of a same firm appears since firms are repeatedly entered in the risk set (due to the time varying nature of the covariates) and can experience several spells. The solution to that problem of non-independence has been solved by the estimation of a robust covariance matrix according to the methods developed by Lin and Wei (1989)[29].

4.1   ESTIMATION OF THE OBSOLESCENCE RATE OF THE TECHNOLOGICAL KNOWLEDGE Δ

We have performed a scanning in order to find the value of δ leading to the maximum log-likelihood for the Model 1. δ is a parameter of particular importance since it provides an indication of the rate at which the relevance of previously accumulated technological knowledge decreases for the prediction of firms' probability of persistence in innovation. If $\delta^*=1$ it means that the past technological activity should not be taken into account in the stock of technological knowledge relevant for the prediction of the probability of persistence in innovation. On the contrary, if $\delta^*=0$ it means that current and past technological knowledge should be considered as equal contributors to the stock of technological knowledge relevant for the prediction of the level of persistence in innovation.

Firstly, we have supposed that δ was similar in France and in the UK ($\delta_F=\delta_{UK}=\delta$) Table 4 reports the likelihood associated to different value of δ resulting from the estimation of the model 1 including both France and the UK simultaneously[30]. The best estimation is obtained with a high value of δ ($\delta^*\approx0.6$). It indicates that the best predictor of the probability of persistence in innovation is obtained when we consider that the past technological knowledge contributing to the current stock of technological knowledge looses each year approximately 60% of its efficiency.

When model 1 is estimated separately in each country, the best fit is obtained with a slightly larger value of δ^* in the UK ($\delta^*_{UK}\approx0.625$) than in France ($\delta^*_F\approx0.575$). It indicates that the stock of technological knowledge developed through the past patenting activity and relevant for the development of subsequent patents tends to depreciate faster in the UK than in France. Descriptive statistics concerning K_t are reported in Appendix 2 (for δ=0.6).

Table 4: Log likelihood of the Model 1 depending on the value of δ

	Initial log(L)	0	0.2	0.4	0.55	0.575	0.6
F+UK (δF=δUK)	-6478.6	-5332.4	-5244.6	-5199.6	-5187.0	-5186.3	-5186.0
France only	-3288.9	-2565.6	-2518.8	-2495.9	-2490.1	-2489.9	-2490.0
UK only	-3189.6	-2766.8	-2725.7	-2703.7	-2696.9	-2696.4	-2696.0

(Table 4 continued) Log likelihood of the Model 1 depending on the value of δ

	Initial log(L)	0.625	0.65	0.7	0.8	1
F+UK (δF=δUK)	-6478.6	-5186.1	-5186.7	-5189.3	-5200.8	-5257.6
France only	-3288.9	-2490.3	-2490.8	-2492.6	-2499.0	-2528.2
UK only	-3189.6	-2695.8	-2695.9	-2696.7	-2701.8	-2729.3

Graph 13: Log likelihood of the Model 1 depending on the value of δ when it is alternatively estimated with French and UK firms

France   UK

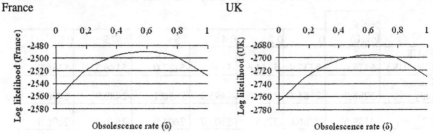

4.2.   PRESENTATION OF THE RESULTS

The results reported below have been obtained with $\delta_F=\delta_{UK}=0.6$ since it generates the best fit when French and UK data are simultaneously used (see Table 3). That constraint of a same obsolescence rate in France and in the UK is a way to neutralize its impact on the estimated coefficients associated to the main variables of interest in our study.

The estimated coefficients in *Table 5* and   *Table 6* report the impact of the covariates on the hazard rate resulting from the estimation of a Cox model stratified by countries and including time varying covariates[31]. In these tables a positive coefficient reflects a positive impact on the probability of interruption of the innovative spell (i.e. a negative impact on the probability of persistence in innovation). A negative coefficient reflects a negative impact on the probability of interruption of the innovative spell (i.e. a positive impact on the probability of persistence in innovation).

Table 5: Results from the estimation of a Cox model with time varying covariates (Model 1)

Endogenous variable t: spell length (criterion for interruption: 1 year without patent). Output from SAS Proc PHREG; $\delta_F=\delta_{UK}=\delta=0.6$; ties handled with the exact method; robust sandwich variance estimates are reported. <u>Different baseline hazard functions are estimated for each country.</u>	**Model 1**		
		France ($_F$)	**UK** ($_{UK}$)
Maximum Likelihood Estimates			
Pre-spell technological experience: lag: [1.. 3]	EXPE$_1$	-0.517* (0.0281)	-0.192* (0.0221)
	EXPE$_2$	-0.225* (0.0534)	-0.076 (0.0437)
lag: [4..6]	EXPE$_3$		
lag>6		-0.178* (0.0902)	-0.018 (0.0754)
Level of technological accumulation	Accu$_t$	-0.57* (0.0389)	-0.476* (0.0297)
Level of technological accumulation squared	Accu$_t^2$	0.004* (0.0003)	0.007* (0.0003)
Interaction of t with the level of technological accumulation	Accu$_t$×t	-0.001 (0.0089)	-0.012* (0.0055)
Model Fit Statistics			
Pseudo R²	R²L	19.95%	
Akaike Information Criterion	AIC	10477.4	
Testing of the Global Null Hypothesis: BETA=0			
Degree of Freedom	d.f.	12	
Likelihood Ratio	L.R.	2585.2*	
Score test	Score	1395.2*	
Wald chi-square	Wald	1640.9*	

* indicates estimated coefficients significant at 5%. In parenthesis: robust standard error of the estimated coefficient that takes account of repeated entry in the risk set and multi-spell firms.

Table 6: Results from the estimation of a Cox model with time varying covariates (Model 2)

Endogenous variable t: spell length (criterion for interruption: 1 year without patent). Output from SAS Proc PHREG; $\delta_F=\delta_{UK}=\delta=0.6$; ties are handled with an exact method; robust sandwich variance estimates are reported. <u>Different baseline hazard functions are estimated for each country.</u>	Model 2	
	France ($_F$)	UK ($_{UK}$)
Maximum Likelihood Estimates		
Pre-spell technological experience: lag: [1.. 3]	**EXPE$_1$** -0.519* (0.028)	-0.193* (0.0229)
	EXPE$_2$ -0.226* (0.0534)	-0.072 (0.045)
lag: [4..6]	**EXPE$_3$**	
lag>6	-0.174 (0.0899)	-0.01 (0.0768)
Level of technological accumulation: t:[1..3]	**ACCU$_{t1}$** -0.538* (0.0284)	-0.476* (0.0246)
	ACCU$_{t2}$ -0.575* (0.0339)	-0.454* (0.0359)
:[4..6]	**ACCU$_{t3}$**	
t>6	-0.522* (0.0593)	-0.568* (0.0562)
Model Fit Statistics		
Pseudo R²	R²L	19.83%
Akaike Information Criterion	AIC	10412.3
Testing of the Global Null Hypothesis: BETA=0		
Degree of Freedom	d.f.	12
Likelihood Ratio	L.R.	2568.8*
Score test	Score	924.0*
Wald chi-square	Wald	1496.7*

* indicates estimated coefficients significant at 5%. In parenthesis: robust standard error of the estimated coefficient that takes account of repeated entry in the risk set and multi-spell firms.

Impact of the pre-spell technological experience

These two models report the same positive and significant impact of the pre-spell technological experience of the firm on the probability of persistence: the estimated coefficients associated to $EXPE_{1\_F}$ and $EXPE_{1\_UK}$ are negative and significant. In comparison to firms without any previous experience in the patenting activity, firms that have a recent experience (lag≤3) benefit of an impressive $40\%^{32}$ decline of their hazard rate in the French sample and 17.3% in the UK sample. However the longer is the lag between the current spell and the previous spell, the lower is the impact of that past technological experience on the probability of persistence. The loss of efficiency of the past technological experience is quick and significant[33] since the reduction of hazard due to a past technological experience with a lag in-between 4 and 6 years falls to $20\%^{34}$ in the French sample and is no longer significant in the UK one. For a lag of more than 6 years the past technological experience still influences positively the probability of persistence in French sample whereas in the UK one the positive impact tends to zero and is no longer significant. That result confirms the idea already mentioned that the obsolescence rate of the technological knowledge is very fast and certainly larger in the UK sample than in the French one. Interestingly we notice that firms included in the French sample benefit more than UK firms of the positive impact of their past patenting activity as indicated by the significant difference between $EXPE_{i\_F}$ and $EXPE_{i\_UK}$ for $i=1$, 2^{35}.

H1: "technological accumulation stimulates persistence"

The basic hypothesis of our work ($H_1 : \dfrac{\partial h_t}{\partial K_t} < 0 \ \forall t$) seems particularly robust in both samples. Model 1 reports a significant and negative estimated coefficient for $ACCU_t$. Model 2 confirms that analysis since the estimated coefficients associated to $ACCU_{ti}$ and for $i=1$, 2, 3 are all negative and significant. It means that whatever the time already spent in the patenting activity, an increased level of technological accumulation induces always an increase of the probability of persistence in the patenting activity.

We notice that the impact of the level of technological accumulation on the hazard rate tends to be larger in the French sample than in the UK one[36]; that difference would be mainly marked for intermediate duration time (when the spell length is in-between 4 to 6 years) as indicated by the significant difference between $ACCU_{t2\_F}$ and $ACCU_{t2\_UK}{}^{37}$ in model 2.

H2: "technological accumulation stimulates persistence at a decreasing rate"

$ACCU_t^2$ aims at grasping an eventual non-linear impact of the level of technological accumulation on the probability of persistence. The positive and significant estimated coefficients associated to that variable both samples indicate that the hypothesis $H_2 : \dfrac{\partial^2 h_t}{\partial K_t^2} > 0$ $\forall t$ should not be rejected. Combined with the previous result concerning H_1, it means that when K_t increases, the probability of persistence does increase for low values of K_t but at a decreasing rate up to a point (at $K_t \approx 68$ in France and 35 in the UK[38]) after which any further technological accumulation leads to a lower probability of persistence. That decreasing impact is significantly larger in the UK sample than in French one[39]. Following our theoretical analysis it may indicate that firms included in the French sample manage more easily than firms of the UK sample the trade-off between exploration and exploitation strategies.

H3: "an increasing impact of the technological accumulation through the spell"
    The last major hypothesis of our work deals with the increasing impact of the level of technological accumulation on the probability of persistence in the course of the spell ($H_3 : \dfrac{\partial}{\partial t}\left(\dfrac{\partial h_t}{\partial K_t}\right) < 0$ $\forall t$). That hypothesis is investigated with two variables:
- an interaction variable between the level of technological accumulation and t (ACCU$_t$×t);

- a split variable in the form of ACCU$_{ti}$ (i=1.2.3).

    The results obtained in models 1 and 2 are not concluding.
    In the French sample neither the first nor the second model indicate any time varying impact of the level of technological accumulation on the probability of persistence: in Model 1 ACCU$_{t\_F}$×t is not significant ; in Model 2 ACCU$_{t1}$, ACCU$_{t2}$, and ACCU$_{t3}$ are not statistically different from each other. On the contrary, in the UK sample Model 1 indicates a negative and significant interaction between the level of technological accumulation and the length of the spell (ACCU$_{t\_UK}$×t). Model 2 tends to confirm that phenomenon since it reports an important increase of the impact of the level of technological accumulation on the probability of persistence when spells get longer (more than 6 years). However, the observed differences between ACCU$_{t1\_UK}$, ACCU$_{t2\_UK}$, and ACCU$_{t3\_UK}$ are not statistically significant at the 5% threshold[40].

Robustness of the results

Results reported in that paper seem rather stable. Firstly, the comparison between French and UK samples show that our conclusions are qualitatively robust (signs of the estimated coefficients most of the time similar in both samples). Secondly, different values of the obsolescence rate ranging from 0 to 1 have been used in order to evaluate the sensitivity of the results to that parameter. We observe that the estimated coefficients associated to the covariates remain relatively stable for intermediate values of δ (between 0.4 to 0.8). Thirdly, in the case of France, it was also possible to perform a stratification by main field of technological activity (see Appendix n°3). That change does not lead to important departures from the results reported in *Table 5* for France. Lastly, the robustness of the assumption of partial independence between the successive spells potentially experienced by a firm has been evaluated. As indicated in Appendix 4 that assumption does not induce significant disturbances. In particular, signs of the estimated coefficients are not changed.

5. DISCUSSION

The purpose of that research was to test the existence of a positive link between the level of technological accumulation (measured with patent data granted in the US-PTO) and the ability of firms to persist in that patenting activity. Behind that question lays the problem of the cumulative nature of the technological change and the likely existence of reinforcement / saturation processes leading to the stimulation / exhaustion of the benefits resulting from that accumulation. In contrast with previous econometric analysis of the same datasets (Geroski et al. (1997) for UK firms and Le Bas et al. (2001) for French firms) we use a semi-parametric Cox model of duration rather than a Weibull model since it rules out any risk of misspecification of the hazard function. In addition we use time varying covariates. The conclusions obtained in that work are coherent with several other empirical studies that, however, use different types of data[41].

The major interest of that survival time analysis is that it clarifies the distinction between the static and dynamic dimensions of the process of technological accumulation. The static dimension refers to the capacity of the firm to use the knowledge related to its past technological activity for the development of subsequent innovations that will make it survivor in innovation. The phenomenon is often considered as the sign of a "learning" process. The dynamic dimension refers to the evolution through time of the efficiency with which the stock of knowledge related to past periods of technological activity is used for the development of subsequent innovations. That phenomenon is supposed to be linked to a process of "learning to learn".

From a static point of view we firstly show that firms' pre-spell technological experience increases strongly their probability of persistence in the patenting behavior provided it is not too old. It is the sign of a short-term memory effect: the technological knowledge gained during previous periods of patenting activity is not completely lost when the firm stops patenting; it can be partly transferred even if the efficiency of that reallocation decreases quickly when the lag between the current and the previous periods of patenting activity increases. That analysis is confirmed by the high rate of obsolescence of the technological knowledge for which the likelihood of the model is maximized ($\delta \approx 60\%$). Secondly we observe both in France and in the UK, that the ability of new innovators to keep on patenting is strongly and positively influenced by the level of technological accumulation they have made during the current spell (H_1). From our point of view it indicates that a learning process takes place which is linked to the accumulation of related technological knowledge. However, the positive impact of the level of technological accumulation on persistence is not linear. When firms accumulate technological knowledge, a saturation process seems to take place (H_2): the marginal contribution of each new patent to the probability of persistence is decreasing. It may reflect difficulties in the management of the trade-off between exploration and exploitation of new technologies.

From a dynamic point of view we do not obtain strong conclusions indicating that the impact of the level of technological accumulation on the probability of persistence increases when spells get longer: no significant relationship is observed in the French sample whereas in the UK sample the relationship is not robust. However that result is very important since it points out at least that the impact of the level of technological accumulation on the probability of persistence does not decrease during the spell. It means that the ability of the firms to learn from their past patenting activity in order to develop new patents is certainly stable through time.

ENDNOTES

[1] The analysis is carried out with data about patents granted by the US Patent and Trademark Office to French and UK firms over the 1971-1984 period. We use the same datasets as Geroski, Van Reenen and Walters (1997) for the UK and Le Bas, Cabagnols and Gay (2001).

[2] That specific type of indicators has been made available through European CIS surveys. See Barlet, Duguet, Encaoua and Pradel (1998).

[3] In that paper, a spell will refer to a non-stop period of patenting activity in the US-PTO.

[4] Our work explores that question with patent data. Consequently, we will only be interested in potential feed-backs coming from and going to the development of patented technologies in the US-PTO.

[5] "Other factors" refer in particular to the economic success of the innovative behaviour and to the level of technological diversity of the firm.

[6] Firm size may follow the same kind of process: an initial reinforcement process (explained by an easier access to financing, higher market power) that is linked to the appearance of an inhibiting phenomenon (bureaucracy, coordination problems ...) so that firm size finally reaches a threshold (Cohen and Klepper (1996)). Learning by doing can also be mentioned in a different way since it is a depletion process of a-priori fixed opportunities of technological improvement.

[7] Econometric investigations have been carried out with two statistical packages: SAS 9.2 (proc PHREG) and Stata 6 (command STCOX). Both allow for: time dependent covariates; censoring; repeated failure and correction for correlations between observations of a same individual in the form of robust variance estimations.

[8] That problem may be at least partially overcome if it was possible to determine with accuracy the level of sensitivity and specificity of such patent indictors of innovation.

[9] See Geroski et al. (1997), p.35 footnote n°3.

[10] From an econometric point of view it is technically possible to handle situations where the dependent variable is left censored. However, in our case, when the dependent variable is left-censored, independent variables can't be accurately defined.

[11] Geroski et al. (1997) do not perform that control whereas they use as an explanatory variable the "number of patents granted at the start of the spell". Since left censored observations were included measurement errors inevitably occurred. In particular, using their dataset, we notice that the mean number of patents granted at the start of the spell to left censored observations is on average 3.37, whereas it is only 1.30 for non left-censored observations. We also notice that whereas only 12% of the observations of their database are left censored, they account for 68% of the spells starting with 5 patents or more. It indicates that left censored observations are

very different from non-left-censored ones; many of them have certainly entered into innovation many years before the start of the study, which may explain why they patent so much.

[12] For France and UK right-censored spells (with no clearly defined ending date) are those whose last observation is made in 1984.

[13] Log-rank test: chi2(1)= 0.24, Pr>chi2=0.62. Wilcoxon test: chi2(1) =2.42, Pr>chi2 =0.12.

[14] Unobserved heterogeneity arises when we aggregate individuals following different hazard functions without being able to identify the discriminating factor. In the simplest situation, if we aggregate several populations following different constant hazard functions, we obtain mechanically for the whole sample a decreasing hazard function (Gourieroux (1989)).

[15] Usually, estimating firstly a generalized gamma model does the discrimination between these different models (Kalbfleisch and Prentice (1980), p.63). That procedure has not been applied here since we have encountered serious convergence problems (faced for UK and France) in the estimation of the generalized gamma model.

[16] Other specific functions could have been applied. However, if they are not nested they can't be easily compared using the likelihood ratio or the Wald test.

[17] It is also used by Geroski et al. (1997).

[18] "observed" refers to Kaplan-Meier estimates.

[19] Its is the same type of hazard function as those obtained by Geroski, Van Reenen and Walters (1997) for UK and Le Bas and Cabagnols (2001) for France.

[20] See Kalbfleish and Prentice (1980), p.67 for a discussion concerning the problems met when the functional form is loosely specified.

[21] Notice that the baseline hazard/survivor function obtained with a Cox model without covariates is similar to that obtained with the Kaplan-Meier method. See Kalbfleisch and Prentice (1980), chapter 4.

[22] The PH hypothesis states that "the hazard ratio comparing any two specifications of predictors is constant over time" or similarly that ratio between the hazard function of one individual to the hazard function of an other individual is constant over time (Kleinbaum (1996)).

[23] Variables remaining fixed during the whole period when subjects are at risk.

[24] That dependence with time can be mechanical, dependent on the behaviour of firms or ancillary if linked to the context rather than to the behaviour of the individuals under study.

[25] See for example Hall and Mairesse (1995) and Crépon and Duguet (1997).

[26] We consider implicitly that each patent represents an equal increase of the technological knowledge, neglecting differences in patent value for a same firm,

between sectors and countries. Moreover, we ignore any increase in the stock of technological knowledge that does not result in a patenting activity in the US PTO. Clearly, what is measured is the approximate level of technological accumulation embodied in patents granted by the US PTO.

[27] No objective depreciation rate exists for the physical capital stock. For intangible assets the problem is still more difficult. We examine this issue latter in the paper.

[28] Since our observation time starts in 1969, we under evaluate the number of firms with a previous technological experience when that last experience is recorded before that date. The probability of such a measurement error decreases when the starting date of the spell increases.

[29] That method is implemented in SAS and Stata (see StataCorp (2001) instructions for STCOX and SAS (2004) instructions for PHREG).

[30] Similar results are obtained with Model 2.

[31] Time is measured in a discrete way. Covariates are measured at each time t. The hazard rate in t is the probability to exit innovation in time t+1 conditionally on the fact that the firm was still patenting in time t.

[32] $0.65=1-\exp(-0.517)$ where -0.517 is the estimated coefficient associated to $EXPE_{1\_F}$ in model 1.

[33] In model 1 the test of the hypothesis H_0: $EXPE_{1\_F}=EXPE_{2\_F}$ leads to a Wald $Chi^2=27.61$; with a probability $P(Chi^2(1df)>27.61)<0.0001$ which induces the rejection of H_0 and the conclusion that for firms included in the French sample the increase in persistence resulting from a recent technological experience (less than 4 years hold) is significantly larger than the increase resulting from an older technological experience (taking place in-between 4 and 6 years ago). The same conclusion holds for the UK sample (H_0: $EXPE_{1\_UK}=EXPE_{2\_UK}$ leads to a Wald $Chi^2=6.7$ with $P(Chi^2(1df)>6.7)=0.0096$).

[34] $0.184=1-\exp(-0.203)$ where -0.203 is the estimated coefficient associated to $EXEP_{2\_F}$ in model 1.

[35] For example in model 1 the test of the hypothesis H_0: $EXPE_{1\_F}=EXPE_{1\_UK}$ leads to a Wald $Chi^2=82.5$; with a probability $P(Chi^2(1df)>82.5)<0.0001$ which induces the rejection of H_0 and leads to the conclusion that the increased persistence resulting from a recent technological experience (less than 4 years) is significantly larger in the French sample than in the UK one. The difference between $EXPE_{2\_F}$ and $EXPE_{2\_UK}$ is also significant at the 5% threshold. For older technological experiences ($EXPE_3$) we do not notice any significant difference between samples.

[36] In model 1 the test of the hypothesis H_0: $ACCU_{t\_F}=ACCU_{t\_UK}$ leads to a Wald chi-square of 3.67 above which we fall in only 5.5% of the cases if H_0 is true.

[37] In model 2, H_0: $ACCU_{ti\_F}=ACCU_{ti\_UK}$ is rejected at the 5% threshold for $i=2$ (p value=0.014) but can't be rejected for $i=1$ (p value=0.09) and $i=3$ (p value=0.58).

[38] These figures must be compared to the structure of the sample: in France the maximum level of technological accumulation K_t ever reached by a firm is 129.6 (the

mean is 3.09 for persistent firms), in the UK the maximum is 67.2 patents (the mean is 2.7 for persistent firms). It means that a very few firms reach a sufficient level of technological accumulation to enter in a situation where any further technological accumulation leads to a decreased probability of persistence in comparison to their former situation.

[39] In model 1 we have tested: H_0: $ACCU_t^2{}_F = ACCU_t^2{}_{UK}$ which leads to a Wald $Chi^2 = 40.28$; with a probability $P(Chi^2(1df)>40.28)<0.0001$. At the 5% threshold H_0 can be rejected.

[40] H_0: $ACCU_{t3\_UK} = ACCU_{t2\_UK}$ leads to a Wald Chi^2 of 3.17 with a probability $P(Chi^2(1df)>3.17)<0.074$; H_0: $ACCU_{t3\_UK} = ACCU_{t1\_UK}$ leads to a Wald Chi^2 of 2.67 with a probability $P(Chi^2(1df)>2.67)<0.1$.

[41] Crépon and Duguet (1997); Cefis and Orsenigo (2001) with European Patent Office data; Gruber (1992) and (1994) with different case studies of the semi-conductor industry; Cabagnols (2000); Duguet and Monjon (2002) with French survey data; Martinez-Ross and Labeaga (2002) with Spanish survey data.

APPENDIX 1: KAPLAN-MEIER ESTIMATE OF THE SURVIVOR FUNCTION FOR DIFFERENT VALUES OF THE NUMBER OF PATENTS AT THE START OF THE SPELL IN THE UK (1969-88)

Table 7a. Kaplan-Meier estimates of the survivor function for different values of the number of patents at the start of the spell in the UK: One patent exactly at the start of the spell (patl)[*]

t	Total	Fail	Lost	Survivor Function s(t)	Std Error	[95%	Conf Inf]	Hazard rate h(t)
1	4036	3038	147	0.2473	0.0068	0.2341	0.2607	0.7527
2	851	479	21	0.1081	0.0051	0.0983	0.1184	0.5629
3	351	163	16	0.0579	0.004	0.0504	0.0660	0.4644
4	172	67	4	0.0353	0.0032	0.0294	0.0421	0.3895
5	101	33	7	0.0238	0.0027	0.0189	0.0296	0.3267
6	61	16	1	0.0176	0.0024	0.0133	0.0228	0.2623
7	44	11	4	0.0132	0.0021	0.0094	0.0179	0.2500
8	29	7	0	0.0100	0.0019	0.0067	0.0144	0.2414
9	22	4	2	0.0082	0.0018	0.0052	0.0123	0.1818
10	16	4	0	0.0061	0.0016	0.0036	0.0100	0.2500
11	12	0	1	0.0061	0.0016	0.0036	0.0100	-
12	11	0	1	0.0061	0.0016	0.0036	0.0100	-
13	10	1	1	0.0055	0.0016	0.0031	0.0093	0.1000
14	8	2	0	0.0041	0.0014	0.002	0.0079	0.2500
16	6	2	0	0.0028	0.0012	0.0011	0.0063	0.3333
18	4	1	2	0.0021	0.0011	0.0007	0.0054	0.2500
20	1	0	1	0.0021	0.0011	0.0007	0.0054	-

[*] Data from Geroski et al. (1997) – 1969-88

Table 7b. Kaplan-Meier estimates of the survivor function for different values of the number of patents at the start of the spell in the UK: Two patents exactly at the start of the spell (pat2)[*]

t	Total	Fail	Lost	Survivor Function s(t)	Std Error	[95%	Conf Int]	Hazard rate h(t)
1	721	388	24	0.4619	0.0186	0.4251	0.4978	0.5381
2	309	136	13	0.2586	0.0167	0.2265	0.2917	0.4401
3	160	64	5	0.1551	0.0142	0.1286	0.1840	0.4000
4	91	31	2	0.1023	0.0121	0.0801	0.1275	0.3407
5	58	13	5	0.0794	0.0109	0.0597	0.1025	0.2241
6	40	9	4	0.0615	0.0100	0.0439	0.0830	0.2250
7	27	5	2	0.0501	0.0093	0.034	0.0707	0.1852
8	20	1	0	0.0476	0.0092	0.0318	0.0680	0.0500
9	19	1	1	0.0451	0.0090	0.0297	0.0652	0.0526
10	17	1	0	0.0425	0.0089	0.0274	0.0624	0.0588
11	16	5	1	0.0292	0.0078	0.0165	0.0476	0.3125
12	10	2	0	0.0233	0.0073	0.012	0.0411	0.2000
13	8	1	0	0.0204	0.0069	0.0099	0.0377	0.1250
14	7	1	0	0.0175	0.0065	0.0079	0.0342	0.1429
15	6	0	1	0.0175	0.0065	0.0079	0.0342	-
16	5	1	0	0.0140	0.0061	0.0055	0.0303	0.2000
17	4	1	0	0.0105	0.0055	0.0033	0.0263	0.2500

[*] Data from Geroski et al. (1997) – 1969-88

Table 7c. Kaplan-Meier estimates of the survivor function for different values of the number of patents at the start of the spell in the UK: Three patents exactly at the start of the spell (pat3)[*]

t	Total	Fail	Lost	Survivor Function s(t)	Std Error	[95%	Conf Inf]	Hazard rate h(t)
1	224	82	3	0.6339	0.0322	0.5672	0.6932	**0.3661**
2	139	52	3	0.3968	0.0329	0.3322	0.4606	**0.3741**
3	84	25	3	0.2787	0.0304	0.2208	0.3394	**0.2976**
4	56	12	5	0.2190	0.0284	0.1661	0.2767	**0.2143**
5	39	9	3	0.1684	0.0264	0.1205	0.2233	**0.2308**
6	27	2	1	0.1560	0.0258	0.1094	0.2101	**0.0741**
7	24	4	0	0.1300	0.0246	0.0865	0.1824	**0.1667**
8	20	0	1	0.1300	0.0246	0.0865	0.1824	-
9	19	3	1	0.1094	0.0234	0.0690	0.1603	**0.1579**
10	15	1	0	0.1021	0.0229	0.0629	0.1524	**0.0667**
11	14	4	1	0.0730	0.0205	0.0395	0.1199	**0.2857**
13	9	1	0	0.0649	0.0198	0.0333	0.1109	**0.1111**
14	8	3	0	0.0405	0.0166	0.0163	0.0825	**0.3750**
16	5	1	0	0.0324	0.0151	0.0114	0.0724	**0.2000**
17	4	1	0	0.0243	0.0133	0.0070	0.0619	**0.2500**
19	3	1	0	0.0162	0.0111	0.0033	0.0509	**0.3333**
20	2	0	2	0.0162	0.0111	0.0033	0.0509	-

[*] Data from Geroski et al. (1997) – 1969-88

Table 7d. Kaplan-Meier estimates of the survivor function for different values of the number of patents at the start of the spell in the UK: Four patents exactly at the start of the spell (pat4)[*]

t	Total	Fail	Lost	Survivor Function s(t)	Std Error	[95%	Conf Inf]	Hazard rate h(t)
1	87	24	1	0.7241	0.0479	0.6173	0.8058	**0.2759**
2	62	15	1	0.5489	0.0536	0.4382	0.6466	**0.2419**
3	46	15	0	0.3699	0.0524	0.2686	0.4713	**0.3261**
4	31	5	1	0.3103	0.0503	0.2154	0.4098	**0.1613**
5	25	4	2	0.2606	0.0480	0.1723	0.3576	**0.1600**
6	19	1	1	0.2469	0.0474	0.1604	0.3433	**0.0526**
7	17	3	0	0.2033	0.0452	0.1232	0.2977	**0.1765**
10	14	1	1	0.1888	0.0442	0.1113	0.2821	**0.0714**
11	12	3	0	0.1416	0.0407	0.0737	0.2311	**0.2500**
13	9	1	0	0.1259	0.0391	0.0620	0.2134	**0.1111**
14	8	1	1	0.1101	0.0373	0.0508	0.1953	**0.1250**
16	6	2	0	0.0734	0.0327	0.0261	0.1541	**0.3333**
17	4	1	1	0.0551	0.0292	0.0158	0.1319	**0.2500**
18	2	1	0	0.0275	0.0243	0.0030	0.1088	**0.5000**
20	1	0	1	0.0275	0.0243	0.0030	0.1088	-

[*] Data from Geroski et al. (1997) – 1969-88

Table 7e. Kaplan-Meier estimates of the survivor function for different values of the number of patents at the start of the spell in the UK: five patents exactly at the start of the spell (pat5)[*]

t	Total	Fail	Lost	Survivor Function s(t)	Std Error	[95%	Conf Inf]	Hazard rate h(t)
1	110	12	2	0.8909	0.0297	0.8159	0.9365	0.1091
2	96	8	2	0.8167	0.0371	0.7304	0.8776	0.0833
3	86	8	0	0.7407	0.0422	0.6469	0.8132	0.0930
4	78	8	0	0.6647	0.0457	0.5667	0.7455	0.1026
5	70	8	4	0.5888	0.0477	0.4893	0.6753	0.1143
6	58	4	0	0.5482	0.0485	0.4483	0.6373	0.0690
7	54	4	1	0.5076	0.0490	0.4080	0.5987	0.0741
8	49	3	0	0.4765	0.0492	0.3776	0.5688	0.0612
9	46	4	0	0.4350	0.0491	0.3377	0.5283	0.0870
10	42	3	0	0.4040	0.0487	0.3083	0.4974	0.0714
11	39	5	0	0.3522	0.0477	0.2604	0.4451	0.1282
12	34	1	1	0.3418	0.0474	0.2509	0.4346	0.0294
13	32	1	0	0.3311	0.0471	0.2412	0.4236	0.0313
14	31	2	0	0.3098	0.0464	0.2219	0.4017	0.0645
15	29	3	0	0.2777	0.0451	0.1934	0.3682	0.1034
16	26	4	1	0.2350	0.0430	0.1565	0.3228	0.1538
17	21	4	0	0.1902	0.0402	0.1189	0.2744	0.1905
18	17	1	0	0.1790	0.0394	0.1097	0.2621	0.0588
20	16	0	16	0.1790	0.0394	0.1097	0.2621	-

[*] Data from Geroski et al. (1997) – 1969-88

APPENDIX 2: DESCRIPTIVE STATISTICS

Table 8a. Evolution of K_t during spells of length 1- 5

Spell Length t	Type of observation	Nb. of spells lasting at least t years n_t		Nb. of patent granted to firms in the tth year of their spell k_t		Stock of technological knowledge accumulated during the spell (δ=0,6) K_t		k_t/n_t		K_t/n_t	
		F	UK	F	UK	F	UK	F	UK	F	UK
1	Failure	2734	2489	3106	2948	3106.0	2948.0	1.1	1.2	1.1	1.2
1	Survivor	785	928	1369	1558	1369.0	1558.0	1.7	1.7	1.7	1.7
1	Censored	313	167	388	198	388.0	198.0	1.2	1.2	1.2	1.2
2	Failure	364	412	496	568	703.2	808.4	1.4	1.4	1.9	2.0
2	Survivor	361	397	942	944	1237.6	1250.0	2.6	2.4	3.4	3.1
2	Censored	60	40	132	56	176.8	76.0	2.2	1.4	2.9	1.9
3	Failure	131	171	220	243	354.5	405.4	1.7	1.4	2.7	2.4
3	Survivor	204	199	753	683	1063.7	994.9	3.7	3.4	5.2	5.0
3	Censored	26	4	100	4	149.8	7.2	3.8	1.0	5.8	1.8
4	Failure	52	65	86	131	151.4	210.9	1.7	2.0	2.9	3.2
4	Survivor	136	116	633	456	961.1	733.0	4.7	3.9	7.1	6.3
4	Censored	16	4	49	8	81.0	13.6	3.1	2.0	5.1	3.4
5	Failure	22	27	37	44	76.4	85.7	1.7	1.6	3.5	3.2
5	Survivor	100	78	485	415	804.9	651.4	4.9	5.3	8.0	8.4
5	Censored	14	4	36	11	61.0	17.4	2.6	2.8	4.4	4.4

Table 8b. Evolution of K_t during spells of length 6- 10

Spell Length t	Type of observation	Nb. of spells lasting at least t years		Nb. of patent granted to firms in the tth year of their spell		Stock of technological knowledge accumulated during the spell (δ=0,6)		kt /nt		Kt/nt	
		nt F	UK	kt F	UK	Kt F	UK	F	UK	F	UK
6	Failure	19	18	31	25	71.1	53.2	1.6	1.4	3.7	3.0
6	Survivor	76	54	373	273	645.1	479.4	4.9	5.1	8.5	8.9
6	Censored	5		15		24.7		3.0		4.9	
7	Failure	13	8	13	14	35.4	34.9	1.0	1.8	2.7	4.4
7	Survivor	51	43	317	229	527.2	391.8	6.2	5.3	10.3	9.1
7	Censored	12	1	41	1	66.5	2.0	3.4	1.0	5.5	2.0
8	Failure	7	8	17	17	28.5	32.3	2.4	2.1	4.1	4.0
8	Survivor	39	31	249	189	430.8	312.5	6.4	6.1	11.0	10.1
8	Censored	5		40		57.6		8.0		11.5	
9	Failure	5	5	6	7	15.5	14.8	1.2	1.4	3.1	3.0
9	Survivor	31	24	250	145	407.7	258.4	8.1	6.0	13.2	10.8
9	Censored	3		11		16.1		3.7		5.4	
10	Failure	4	2	10	2	19.0	4.1	2.5	1.0	4.8	2.0
10	Survivor	18	19	127	107	213.8	201.3	7.1	5.6	11.9	10.6
10	Censored	9	1	69	2	136.3	3.6	7.7	2.0	15.1	3.6

Table 8c. Evolution of K_t during spells of length 11-14

Spell Length t	Type of observation	Nb. of spells lasting at least t years nt		Nb. of patent granted to firms in the tth year of their spell kt		Stock of technological knowledge accumulated during the spell (δ=0,6) Kt		kt /nt		Kt/nt	
		F	UK	F	UK	F	UK	F	UK	F	UK
11	Failure	1	4	9	6	24.0	12.0	9.0	1.5	24.0	3.0
11	Survivor	13	12	77	92	134.1	157.9	5.9	7.7	10.3	13.2
11	Censored	4	2	22	2	35.4	6.8	5.5	1.0	8.8	3.4
12	Failure	0	1		2		3.8		2.0		3.8
12	Survivor	6	9	50	62	79.3	113.6	8.3	6.9	13.2	12.6
12	Censored	7	0	20		44.3		2.9		6.3	
13	Failure	0	1		1		1.9		1.0		1.9
13	Survivor	3	7	24	30	42.9	64.3	8.0	4.3	14.3	9.2
13	Censored	3	0	13		25.8		4.3		8.6	
14	Failure	0	0								
14	Survivor	0	4		17		30.8		4.3		7.7
14	Censored	3	0	46		63.2		15.3		21.1	

Appendix 3: Stratification by main field of technological activity

Table 9: Results from the estimation of a Cox model with time varying covariates and stratification by main field of technological activity (France only)

Endogenous variable t: spell length (criterion for interruption: 1 year without patent). Output from SAS Proc PHREG; δ_F =0.6; ties handled with the exact method; robust sandwich variance estimates are reported. Different baseline hazard functions are estimated for each technological field.		Model 1 France
Maximum Likelihood Estimates		
Pre-spell technological experience: lag: [1.. 3] lag: [4..6] lag>6	EXPE₁	-0.525* (0.0291)
	EXPE₂	-0.208* (0.0566)
	EXPE₃	-0.206* (0.0936)
Level of technological accumulation	ACCU$_t$	-0.57* (0.0396)
Level of technological accumulation squared	ACCU$_t^2$	0.004* (0.0003)
Interaction of t with the level of technological accumulation	ACCU$_t$×t	-0.001 (0.009)
Model Fit Statistics		
Pseudo R²	R²L	24.7%
Akaike Information Criterion	AIC	4699.9
Testing of the Global Null Hypothesis: BETA=0		
Degree of Freedom	d.f.	6
Likelihood Ratio	L.R.	1537.3*
Score test	Score	839.3*
Wald chi-square	Wald	907.6*

* indicates estimated coefficients significant at 5%. In parenthesis: robust standard error of the estimated coefficient that takes account of repeated entry in the risk set and multi-spell firms.

A classification with 6 major fields of technological activity is used. The main field of technological activity of a firm is the field in which it has accumulated the maximum level of technological knowledge during the spell.

APPENDIX 4: SENSITIVITY OF THE ESTIMATION TO THE RANK OF THE SPELL

For multi-spell firms, the maximum number of spell observed on the 1971-84 period is 6). We have decided to group together all the spells with a rank superior or equal to 4.

Table 10: Distribution of the spells by rank

Rank of the spell	Number of spells		
	France	UK	Total
1	2711	2044	4755
2	746	983	1729
3	255	375	630
4	87	130	217
5	28	43	71
6	5	9	14
Total	3832	3584	7416

The econometric models estimated in that paper are based on the assumption that a new period of technological accumulation starts at each spell. The impact of previously experienced spells being simply captured with a fixed-effect related to the length of the gap between the current and previous spell (less than 4 years, 4 years to 6 years, more than 6 years i.e. $EXPE_1$, ..., $_3$). That assumption leads to the estimation of only one parameter for each explanatory variable whatever the rank of the spell in which the firm is engaged. In order to test that assumption we have estimated a special version of the Model 1 in which different values of the covariates are estimated for each spell rank. Result of the estimation made simultaneously for all the coefficients is reported in Table 11.

Table 11: Estimation of the model 1 for different rank of the spell

Endogenous variable t: spell length (criterion for interruption: 1 year without patent). Output from SAS Proc PHREG; δ =0.6; ties handled with the exact method; robust sandwich variance estimates are reported. Different baseline hazard functions are estimated for each country.		Rank of the spell			
		Rank =1	Rank =2	Rank =3	Rank ≥4
Estimated parameters for France					
Level of technological accumulation	ACCU$_{t\_F}$	-0.624* (0.0518)	-0.58* (0.0568)	-0.59* (0.0733)	-0.633* (0.1221)
Level of technological accumulation squared	ACCU$_t^2{}_F$	0.004* (0.0004)	0.022* (0.0026)	0.021 (0.0117)	-0.028 (0.0366)
Interaction of t with the level of technological accumulation	ACCU$_{t\_F}$×t	0.004 (0.0138)	-0.013 (0.009)	-0.02 (0.0245)	0.088 (0.0552)
Pre-spell technological experience: lag: [1.. 3]	EXPE$_1$	-0.578* (0.0828)			
	EXPE$_2$	-0.289* (0.0926)			
lag: [4..6]	EXPE$_3$				
lag>6		-0.239* (0.1141)			
Estimated parameters for the UK					
Level of technological accumulation	ACCU$_{t\_K}$	-0.551* (0.0414)	-0.417* (0.0603)	-0.51* (0.0663)	-0.472* (0.1004)
Level of technological accumulation squared	ACCU$_t^2{}_{UK}$	0.008* (0.0005)	0.008 (0.0135)	0.01* (0.0011)	0.035 (0.0211)
Interaction of t with the level of technological accumulation	ACCU$_{t\_UK}$×t	-0.011 (0.0076)	-0.014 (0.0093)	0.007 (0.0127)	0.017 (0.0364)
Pre-spell technological experience: lag: [1.. 3]	EXPE1_UK	-0.377* (0.0827)			
	EXPE2_UK	-0.234* (0.0869)			
lag: [4..6]	EXPE3_UK	-0.18 (0.1021)			
lag>6					
Initial log likelihood	-2logL0	12957.169	d.f. = 30		
Final log likelihood	-2logL	10345.587			

* indicates estimated coefficients significant at 5%. In parenthesis: robust standard error of the estimated coefficient that takes account of repeated entry in the risk set and multi-spell firms.

For each covariate of interest (ACCU$_t$, ACCU$_t^2$ and ACCU$_t$×t) we have tested the null hypothesis H$_0$: "whatever the rank of the spell, the impact of the covariate is similar". If H$_0$ is rejected at the 5% threshold it indicates that the rank matters. Consequently the hypothesis of independence between spells is not realistic. On the contrary, if H_0 is not rejected it means that the hypothesis of independence between spells can not be completely rejected and that the different estimated parameters obtained for each value of the Rank variable can be replaced by only one estimated parameter.

Table 12 reports the results for each variable of interest. In only one case H_0 is rejected (for the estimated coefficient associated to **Accu$_t^2$_F**). Such a result indicates that the hypothesis of quasi independence between spells (in term of value of the estimated parameters) is reasonably admissible.

Table 12: Test of the impact of the rank of the spell on the estimated coefficients

	Wald Chi-Square	DF	Pr > ChiSq	Conclusion concerning HO
Test 1 (ACCUt_F) : H0:ACCUt_F(rank=1)= ACCUt_F(rank=2)=ACCUt_F(rank=3) =ACCUt_F(rank=4)	0.56	3	0.90	Not rejected
Test 2 (ACCUt_UK): H0:ACCUt_UK(rank=1) =ACCUt_UK(rank=2)=ACCUt_UK(rank=3) =ACCUt_UK(rank=4)	6.78	3	0.07	Not rejected
Test 3 (ACCUt² _F): H0:ACCUt² _F(rank=1) =ACCU²t_F(rank=2)=ACCU²t_F(rank=3) =ACCU²t_F(rank=4)	50.87	3	<0.0001	Rejected
Test 4 (ACCU²t_UK): H0:ACCU²t_UK(rank=1)= ACCU²t_UK(rank=2)=ACCU²t_UK(rank=3) =ACCU²t_UK(rank=4)	3.25	3	0.35	Not rejected
Test 5 (ACCUt_Fxt): H0:ACCUt_Fxt(rank=1) =ACCUt_Fxt(rank=2)=ACCUt_Fxt(rank=3) =ACCUt_Fxt(rank=4)	4.31	3	0.22	Not rejected
Test 6 (ACCUt_UKxt): H0:ACCUt_UKxt(rank=1) =ACCUt_UKxt(rank=2)=ACCUt_UKxt(rank=3) =ACCUt_UKxt(rank=4)	2.26	3	0.51	Not rejected

REFERENCES

Barlet, C., Duguet, E., Encaoua, D. and Prade,l J. [1998], "The commercial success of innovations: an Econometric analysis at the firm level in French manufacturing", *Annales d'Economie et de Statistique*, n°49-50, pp.457-478

Bosworth, D. And Jobome, G. [1996], "Rivalry, Competition and the Creative-Destruction of Intellectual Property: a New Approach tot the Modeling of Firm Performance", Presentation for the conference of the Applied Econometric Association (AEA), Luxembourg, November.

Bresch,i S., Malerba, F. and Orsenigo, L. [2000], "Technological Regimes and Schumpeterian Patterns of Innovation", *The Economic Journal*, vol.110, April, pp.388-410.

Cabagnols, A. [2000], "Les déterminants des types de comportements innovants et de leur persistance : analyse évolutionniste et étude économétrique", Unpublished PhD thesis, University Lyon 2.

Cefis, E. and Orsenigo, L. [2001], "The persistence of innovative activities. A cross-countries and cross-sectors comparative analysis", *Research Policy*, n°30, pp.1139-1158.

Cohen, W. M. and Klepper, S. [1996], "A reprise of size and R&D", *The Economic Journal*, vol.106, July, pp.925-951.

Cohen, W. M. and Levin, R. C. [1989], "Empirical studies of innovation and market structure", *in Handbook of industrial Organization*, vol. II, ed. by Schmalensee R. and Willig R.D., Elsevier Science Publishers B.V., pp.1059-1107.

Cohen, W. M. and Levinthal, D. A. [1990], "Absorptive capacity: a new perspective on learning and innovation", *Administrative Science Quarterly*, vol.35, pp.128-152.

Cohen, W. M., Nelson, R. R. and Walsh J. P. [2000], "Protecting Their Intellectual Assets: Appropriability Conditions and Why U.S. Manufacturing Firms Patent (or Not)", NBER Working Paper n° 7552.

Crepon, B. and Duguet, E. [1997], "Estimating the knowledge production function from patent numbers: GMM on count panel data with multiplicative errors", *Journal of Applied Econometrics*, vol.12(3).

Dosi, G. [1988], "Sources, procedures, and microeconomics effects of innovation", *Journal of Economic Literature*, vol.XXVI, sept., pp.1120-1171.

Duguet, E. and Kabla, I. [1998], "Appropriation strategy and the motivations to use the patent system: an econometric analysis at the firm level in French manufacturing", *Annales d'Economie et de Statistique*, n°49-50, pp.289-327

Duguet, E., Monjon, S. [2002], "Creative destruction and the innovative core: is innovation persistent at the firm level ? An empirical examination from CIS data comparing the propensity score and regression methods", Cahiers de la MSE – EUREQua n°2002.69.

Geroski, P. A. [1995], "Markets for technology: knowledge, innovation and appropriability", in *Handbook of the Economics of Innovation and Technical Change*, ed. by Stoneman P., Basil Blackwell, pp.90-131.

Geroski, P. A., Van Reenen, J. and Walters C. F. [1997], "How persistently do firm innovate?", *Research Policy*, 26, pp.33-48.

Gourieroux, C. [1989], *Économétrie des variables qualitatives*, Economica, Paris

Griliches, Z. [1990], "Patent Statistics as Economic Indicators: A Survey," *Journal of Economic Literature*, Vol. 28 (4) pp. 1661-1707.

Gruber, H. [1992], "Persistence of leadership in product innovation", *Journal of Industrial Economics*, vol.40 (4), pp.359-375.

Gruber, H. [1994], Learning and Strategic Product Innovation: Theory and Evidence for the Semi-conductor Industry, Elsevier Science Publishers.

Hall, B. H. and Mairesse, J. [1995], "Exploring the relationship between R&D and productivity in French manufacturing firms", *Journal of Econometrics*, vol.65, pp.263-293

Harabi, N. [1995], "Appropriability of technological innovations: an empirical analysis", *Research Policy*, 24, pp981-992.

Kalbfleisch, J. D. and Prentice, R. L. [1980], *The Statistical Analysis of Failure Time Data*, John Wiley Sons, U.K.

Kleinbaum, D. G. [1996], *Survival Analysis. A self-learning text*, Springer-Verlag, Berlin Heidelberg New-York

Le Bas, C., Cabagnols, A. and Gay, C. [2001], "How persistently do firms innovate? An evolutionary view"- An empirical application of duration models-" in *Applied Evolutionary Economics: New Empirical Methods and Simulation Techniques*, ed. by Saviotti P. P., Edward Elgar.

Levin, R. C., Klevorick, A. K., Nelson, R. R. and Winter, S. G. [1987], "Appropriating the returns from industrial research and development", *Brooking Papers on Economic Activity*, 3, (special issue), pp.783-831.

Malerba, F. and Orsenigo, L. [1995], "Schumpeterian patterns of innovation", *Cambridge Journal of Economics*, 19, pp.47-65.

Malerba, F. and Orsenigo, L. [1999], "Technological entry, exit and survival: an empirical analysis of patent data", *Research Policy*, vol.28, pp643-660.

Malerba, F. et Orsenigo, L. [1993], "Technological regimes and firm behavior," *Industrial and Corporate Change*, vol.2, no.1, pp.45-71.

Malerba, F., Orsenigo, L. and Peretto P. [1997], "Persistence of innovative activities, sectoral patterns of innovation and international technological specialization", *Journal of Industrial Organization*, vol.15, pp.801-826.

March, J. G. [1991], "Exploration and exploitation in organizational learning", *Organization Science*, vol.2, pp.71-87.

Martinez-Ros, E. and Labeaga, J. M. [2002], "Modeling innovation activities using discrete choice panel data models", in *Innovation and Firm Performance: Econometric Explorations of Survey Data*, ed. by Kleinknecht A. and Mohnen P., Palgrave, pp.150-171.

Nelson, R. R. and Winter, S. G. [1982], *An Evolutionary Theory of Economic Change*, The Belknap Press of Harvard University Press, London

SAS Institute [2004], SAS/STAT© 9.1 *User's Guide*. Cary, NC: SAS Institute Inc.

STATACORP [2001*]*, *Stata Reference Manual*, Volume 3 Q-St, Stata Press.

Van Dijk, M. [2000], "Technological Regimes and Industrial Dynamics: The Evidence from Dutch Manufacturing*"*, *Industrial and Corporate Change*, 9(2), June, pp.173-94

Winter, S. G. [1984], "Schumpeterian competition in alternative technological regimes", *Journal of Economic Behavior and Organization*, vol. 5, no. 3-4, pp.287-320.

Chapter 6

PERSISTENT ADOPTION OF TIME-SAVING PROCESS INNOVATIONS

Nilotpal Das, *Futures, LLC*
James G. Mulligan, *University of Delaware*

1. INTRODUCTION

The previous chapters were concerned with persistent innovation by innovating firms. In this chapter we consider the related topic of persistent adoption of innovations by firms that are not directly involved in the innovation process. In addition to a survey of the literature, we offer empirical evidence of persistent adoption for a specific time-saving *process* innovation: the high-speed detachable chairlifts now used at many, but not all, ski resorts.

Despite several papers that indirectly consider persistent adoption of process innovations, there is no unified treatment of this topic in the literature. In addition, while process innovations can lower an adopting firm's costs and/or increase quality, the literature generally assumes that product innovations result in higher quality while process innovations lower costs. For example, see Weiss (2003) for a recent model of a firm's choice between a quality-enhancing product innovation and a cost-reducing process innovation. Despite the literature's emphasis on process innovations that lower costs, much of U.S. productive capacity provides services, not manufactured goods, and with services an important part of quality is the time that consumers spend waiting for and receiving service.

An innovation that improves service time might even raise the adopting firm's costs. More importantly, as shown by Mulligan and Llinares (2003) and Das and Mulligan (2004), a time-saving process innovation can follow a different diffusion path than that generally found for cost-reducing process innovations. The main point of these two papers is that an innovation that increases speed of service is not likely to appeal to all consumers to the same degree. As a result, early adopters may adopt process innovations to target a specific subset of the market leading to increased horizontal differentiation with the adoption rate of previous non-adopters decreasing following adoption by competitors. Generally, the literature finds the opposite result and attributes it either to a spillover of information from adopters to non-adopters or the effects of cost reductions by rivals on non-adopters' market share.

The effect that innovation has on a firm's quality of service can also explain the nature of persistent adoption. To illustrate this point we include in this chapter evidence of persistent adoption of the high-speed chairlift technology by ski resorts. Our analysis accounts for both the unique characteristics of the technology and the quality of service provided by the ski resorts. A major advantage of high-speed chairlift technology for a study of persistent adoption is that it has not undergone significant modification since its first appearance in the United States in 1981. In addition, its characteristics were well known to the U.S. ski industry early on, given that it had already been introduced previously in Europe and had received extensive coverage in the trade literature. As a result, a diffusion study of chairlift technology is less susceptible to the sort of criticism that the evolutionary diffusion literature has directed at studies based on the assumption of well-informed, profit-maximizing firms.

Unlike potential adopters of most other process innovations, ski resorts compete in either local or national submarkets with different levels of quality. This distinction has implications not only for a ski resort's decision to make an initial adoption of high-speed technology but also for the degree of persistence. In Section 3, we provide evidence suggesting that while ski resorts initially adopted high-speed chairlifts in order to differentiate the quality of their service in local markets, ski resorts in national/international submarket were most likely to increase the proportion of their capacity using high-speed technology. We attribute the persistence of adoption to endogenous quality competition in the national submarket consistent with the general implications of work on endogenous quality by Sutton (1991 and 1998). On the other hand, in local markets there is less of an incentive for persistent adoption due to the heterogeneity of the local skiing population and constraints on skiable acres.

2. THE LITERATURE

While the main focus of this chapter is innovation that improves the speed of a firm's service, we start with a survey of the diffusion literature having implications for persistent adoption of process innovations. Although there is an extensive literature that pertains at least indirectly to persistent adoption, the main issues addressed vary considerably and suggest several ways of organizing this work. We have grouped these papers in four general if not entirely mutually exclusive categories: (a) strategic models of continuous investment in technology, (b) learning-by-doing, (c) endogenous quality and vertical differentiation, and (d) neoclassical models of inter-firm and intra-firm diffusion of specific innovations along with criticism of these models coming from the evolutionary diffusion literature.

2.1 STRATEGIC MODELS OF CONTINUOUS INVESTMENT

There is a growing and extensive literature that acknowledges that innovation is often a continuous process. While papers in this literature have various objectives, expectations about the timing, cost, and importance of subsequent improvements in the initial technology all play a role. A prevailing objective concerns the pattern of multiple adoptions and their impact on market leadership with recent experience in rapid innovation in computer technology as a motivating example in many cases. Depending on the assumptions made, firms may either persist in adopting improvements in the technology or delay further adoption.

For example, Thomas (1999) considers empirically the order of adoption for new technology in the computer disk drive industry. His main conclusion is that large firms and incumbents are more likely to adopt minor improvements in the technology with entrants and small firms more likely to adopt innovations that make existing products obsolete. These results are consistent with work by Klepper (1996) on the product life cycle where entrants and smaller firms are the primary adopters of new products. In Thomas' model entrants have a size disadvantage relative to incumbents regarding existing technology and must adopt the new technologies to gain an advantage. Incumbent firms have an incentive to adopt completely new technologies later given that they have already invested in a viable technology. On the other hand, incremental changes in the technology have an effect on efficiency that favors existing firms, which are generally relatively larger.

Thomas provides empirical evidence that the adoption decision was based on both firm size and the nature of the change in the technology. He found that in the workstation submarket the innovation of 3.5" disk drives made the 5.25" technology obsolete, while the same technological change in the desktop and portable submarkets did not eliminate the older technology as a viable option. Incumbents adopted later in the workstation market, but earlier in the other submarket. As a result, knowledge of the effect that the innovation has on the firm's production process and on the consumers of the firm's product is critical in explaining the adoption process.

Earlier, Farzin, Huisman, and Kort (1998) used a dynamic programming model of multiple technology switches given uncertainty about the speed of arrival of new technologies and the magnitude of improvement in productivity. An important feature of the model is that switching costs are assumed to be sunk in order to capture the effect of technological obsolescence. Their model implicitly rules out the possibility that the firm could use both technologies together. They argue that the literature generally focuses on uncertainty about future market conditions and prices, which they hold constant in their model. Their model suggests that the possibility that a more efficient process technology will appear in the future causing firms to delay adoption of improvements of the existing technology.

Athey and Schmutzler (2001) develop a more general model to consider the possibility that a low-cost or high-quality firm will be able to maintain an initial lead over its competitors through persistent advertising, low-price offers to attract new customers, or investments in product and process innovations which are either cost-reducing or demand-enhancing. They establish general conditions for increased market dominance that they apply to many applications, such as network effects, learning by doing, and endogenous quality. In their model two or more oligopolists compete over time with investment possible in each period. Investments can have small or large marginal effects on next period costs or demand. Firms invest in cumulative cost reduction, product quality, a larger stock of loyal customers, or the number of products offered. Their main objective was the derivation of general conditions for whether or not actions are strategic substitutes or complements. Strategic substitutes imply that the incremental benefit from an action, such as a firm's investment in a process innovation decreases with an increase in a competitor's action. Cost reduction and increased product quality can lead to market dominance with ambiguous welfare effects. Since a leading firm may use continuous investment to maintain its market position, their model does imply a role for persistent adoption in a competitive environment.

2.2 LEARNING-BY-DOING

Several papers incorporating assumptions about learning-by-doing also have implications for persistence depending on how quickly firms learn to use the innovation and the extent of information spillovers. The nature of the learning process varies from paper to paper. In some cases learning-by-doing implies persistence, while in others the learning process may actually impede persistent adoption. Despite these differences, this research has in common an emphasis on cost-reducing as opposed to quality-enhancing process innovations.

Learning can come from the firm's own experience with a current or earlier vintage or from the experiences of competitors. For example, Kapur (1995) models a waiting contest with firms delaying adoption in order to learn from those who do adopt. As additional firms adopt, non-adopters update their information concerning the costs of switching to the new technology. This process results in firms adopting at different time periods even though the firms are identical a priori.

Giovanetti (2001) also provides a model where firms delay adoption of new vintages. In his case, however, firms have a cost advantage that comes from continued use of the same vintage. Since learning-by-doing provides no explicit cost advantage from having used an earlier vintage or from learning from others' experiences, this form of learning-by-doing would actually impede persistent adoptions of improvements in the technology. It is even possible to have a leapfrogging of market leaders who alternate as the firm with the latest technology. To show this, Giovanetti models a duopoly with Bertrand competition and continuous innovation providing a sequence of cost-reducing technology. He offers innovations in PC processors as an example. In his model leapfrogging is a necessary condition for long-run technological improvement with the likelihood of leapfrogging depending on the nature of demand for the products produced by the technology.

Just curious: Has anyone developed a general model that has an adoption function which depends on demand and cost relationships so that, depending upon the parameter values produce simultaneous adoption, discrete leapfrogging, etc.?

In a similar vein Karp and Lee (2001) have a model of technology adoption with learning-by-doing where firms using an existing technology may resist adopting a newer technology that requires learning before the full benefit of the technology is realized, especially when there is a high opportunity cost due to lower profits during the learning time period. As a result, firms that have not made an investment in a technology with learning costs have a lower opportunity cost of adopting a new technology. Karp and Lee consider the diffusion pattern with both myopic and forward-looking firms. In some cases lagging firms may overtake leading firms. In their model a high discount rate makes overtaking more likely as relatively backward firms overtake the relatively more technologically advanced firms. On the other hand, forward-looking firms may upgrade to acquire higher skill levels and persist in upgrading more frequently than firms that are not forward-looking.

By contrast, Cabral and Leiblein (2001) find in the case of semiconductor technology that learning-by-doing coming from experience with the most recent vintage of the technology does give firms an incentive to persist in adopting the next vintage even when controlling for firm size. On the other hand, they find no evidence of spillover effects from much earlier vintages or evidence of regional spillover effects due to competitors' adoptions. Persistent adoption comes from accumulated experience with a cost-saving technology.

Taken as a group these papers suggest that persistent adoption for innovations where learning-by-doing is important depends critically on the nature of the learning process. As was the case for the models surveyed in Section 2.1, particular attention needs to be paid to how the innovation actually affects the production process.

2.3 ENDOGENOUS QUALITY

Although not generally included in the literature on technological diffusion, work on market structure and performance by Sutton (1991 and 1998) has implications for the persistent adoption of quality-enhancing innovations. In order to enter a market, firms in Sutton's model incur a sunk fixed cost which is exogenously determined by the technology for producing the good. This cost is the minimum necessary to be able to produce at minimum efficient scale. Free entry results in zero economic profits in long-run equilibrium if firms compete either in Cournot competition or accommodate entry at a collusive price. In markets where product quality is important to consumers, firms make quality-enhancing investments involving sunk costs. These costs are, however, endogenous and increase for each firm as the market expands over time.

While Sutton's main emphasis is the effect that quality-enhancing investments in advertising (Sutton, 1991) or R and D (Sutton, 1998) have on market concentration as market size increases over time, the theoretical approach has implications for persistence of adoption. To the extent quality is determined by the firm's process technology Sutton's model implies that firms would continue endogenously to adopt quality-enhancing improvements in the technology as the market increases. Improvements in quality can lead to horizontal differentiation if markets are competitive and if firms attempt to avoid direct competition with rivals, but the main point of his work is showing how markets may remain concentrated even with expansion in market size and zero profits in long-run equilibrium.

Competition among existing firms making endogenously determined investments in quality increase the degree of vertical differentiation as market size expands. As long as consumers have a common preference for quality, firms may compete by making quality-enhancing investments. In the strictest version of the Sutton model, these investments are assumed to be both fixed and sunk with little impact on variable costs. The level of quality at a point in time is determined endogenously by consumer preferences for quality, the price of an additional unit of quality and the size of the market. Changes in market size over time are due to increases in population and income per capita. Since the main focus of Sutton's approach is the relationship between market size and the degree of concentration in the market and not specifically on the diffusion pattern of innovations, his model is essentially static. For example, while he acknowledges that incumbency can give a firm an advantage by allowing it to determine a level of quality (and investment in quality) that makes entry unprofitable, his model does not explicitly account for the dynamic nature of investments in quality over time.

In a recent application of Sutton's modeling approach to the supermarket industry, Ellickson (2003) assumes that supermarkets invest in distribution technology that could either lower unit costs or raise the quality of the goods sold by the adopting firm. More efficient distribution networks result in larger stores. Ellickson defines quality as the size of a company's store, which is a proxy for the diversity of products offered and not necessarily the quality of the specific goods sold. According to Ellickson, consumers would be willing to pay higher prices for a larger choice of products. This increase in store-level quality is due to endogenous investments in a process innovation related to the company's computerized inventory system.

According to Ellickson, Sutton's model implies that local markets will remain concentrated as market size increases if consumers value larger stores for the greater variety in products offered. Using data for a cross-section of U.S. supermarkets for the year 1998, he finds that store size is positively correlated with population size in local markets even though the number of stores does not change across markets. He argues that if the innovation in the computerized inventory system only lowers cost, it should be reflected in lower prices for a set of products that are sold nationally. When he finds no statistically significant difference in prices across markets and store size, he concludes that consumers must be willing to pay higher prices associated with the greater diversity of products. No controls were included for the possibility that higher quality stores sell a different mix of products of higher quality.

2.4 NEOCLASSICAL DURATION MODELS

We conclude the review of the literature with an extensive survey of the recent work on inter-firm and intra-firm diffusion of innovations. While the inter-firm diffusion literature considers first adoptions, the intra-firm diffusion literature is more directly related to persistent adoption. The diffusion literature attempts to isolate the factors contributing to the diffusion of specific innovations. Karshenas and Stoneman (1995) provide a detailed survey of the earlier work on inter-firm diffusion, while Battisti and Stoneman (2005) summarize the literature on intra-firm diffusion. The intra-firm literature primarily considers the degree to which a firm uses the innovation at its various outlets. In Section 3 we interpret intra-firm diffusion differently by considering the extent of use of high-speed technology at each ski resort. We argue that, as the extent of use increases at the ski resort, the ski resort's overall level of quality increases. In this way intra-firm diffusion of technology is the same as persistent adoption of a quality-enhancing innovation with implications for increased vertical differentiation.

2.5 INTER-FIRM DIFFUSION.

This literature implicitly assumes that otential adopters evaluate the costs and benefits of adoption based on their situation relative to that of other potential adopters and to factors such as firm size, availability of alternative technologies, and price. In order to quantify the relationship between the adoption rate and a set of exogenous factors influencing the adoption rate, researchers have generally used parametric duration models assuming hazard functions with negative exponential or Weibull distributions.

Although the Cox Proportional Hazards model was used by Levin, Levin, and Meisel (1987) and Cabral and Leiblein, it is not common in this literature. As suggested by Karshenas and Stoneman (2002) and Allison (2001), the signs and level of significance of the coefficients should be essentially the same for the Cox model and a properly specified parametric model with a Weibull distribution. While a major advantage of the Cox model is that it places no restriction on the baseline hazard function, a parametric model indicates how the baseline hazard changes over time. Karshenas and Stoneman (1993), for example, found that the baseline hazard in their case increased over time and argued that this was evidence of an epidemic effect of increased informational flows over time among firms. Researchers generally find that the largest firms are the most likely initial adopters and that the hazard rate exhibits positive duration over time. Other factors influencing the adoption decision have been the price of innovation, the age of the firm, and the number of competitors who have adopted.

Despite their wide-spread use, neoclassical models of diffusion have been subject to attack from researchers favoring an evolutionary approach to diffusion. Metcalfe (1988) provides an earlier survey of this literature. The main criticism rests on the assumption of rational decision-making by fully informed optimizing firms. Following Simon (1972), some researchers question whether firms are more likely to make adoption decisions based on a rule of thumb or local custom instead of global optimization in competition with other optimizing firms. They also question the assumption of an equilibrium model with firms making adoption decisions when specific thresholds have been reached, such as the size of the firm, price of the innovation, or number of rivals who have adopted.

Sarkar (1998) provides a more recent survey of this literature with a comparison to the neoclassical approach. According to Sarkar, evolutionary diffusion advocates have had an impact on recent theoretical neoclassical diffusion models, which have increasingly incorporated expectations and the possibility of continuous innovation. Another contribution of the evolutionary approach is an emphasis on the importance of understanding the exact nature of the innovation's impact on the firm's production process. While we argue in Section 3 that a neoclassical approach to the diffusion of chairlift technology is appropriate, we agree that an understanding of the effect that the technology has on the ski resort's quality of service is essential for explaining both initial and persistent adoption of this technology.

Despite several empirical studies of diffusion, data limitations pose a serious problem. For example, according to Sarkar, "the applicability of duration models has ... been ... limited ... mainly because of limitation of data; to estimate duration models, one ideally needs a data set on complete life histories of the population of potential adopters, as well as the characteristics of a well-defined new technology over a sufficiently long period of time since its inception. Such ideal data sets have been relatively rare, and in particular, disaggregated data on the adoption of new technologies have been scarce" (Sarkar, 156-7). As a result, aggregated data are often used that may not reveal the exact nature of the adopting firm's product and the impact that the innovation has on the firm's productivity.

Aggregation can be problematical for an empirical study of persistent adoption if there is no information concerning the nature of improvements in specific innovations over time. A case in point is a recent paper by Bartoloni and Baussola (2001) who use a logit model and cross-section data of adoptions of process and product innovations by Italian manufacturing firms. They regressed a dichotomous variable indicating whether the firms had adopted an innovation during the time period 1990-1992 on various variables, such as firm size, the skill labor of the labor force, marketing expenditures, output growth, and the number of adopting firms in the firm's sector. As is the case for many empirical papers in this literature, the aggregate nature of the data does not permit a determination of causality for any specific innovation or allow for differences in vintages of similar innovations or the firm's history of adoptions. The data also do not reveal whether the innovation is a product or process innovation or a replacement for an earlier innovation.

While the empirical inter-firm diffusion literature generally has relied on aggregate data across broad classes of innovations and vintages, the theoretical literature has focused on the diffusion of individual innovations. Earlier work developed conditions for showing that a diffusion pattern of unequal adoption dates may result even with otherwise identical firms and identical consumers. See for example, Reinganum (1981 and 1983) and Fudenberg and Tirole (1985). Despite this work there has been little guidance for empirical researchers working with data for firms and consumers that are not a priori identical. Most of the theoretical motivation for individual factors is on a variable-by-variable basis. Both Geroski (2000) and Stoneman (2002) provide surveys of the empirical literature with an emphasis on the expected impact of specific variables.

Recently, Götz (1999) offered a model with a cost-reducing process innovation that makes a clear distinction between the effect of epidemic information spillovers and that of adoptions by competitors on the future adoption rate of non-adopters. He models a monopolistically competitive market with identical consumers having preferences for every variety of a specific good and all firms producing with identical production costs a priori. While the new technology lowers the production costs for any firm that adopts, not all firms adopt at the same time. The diffusion pattern is a function of the number of different varieties of the good, the cross-price elasticities, and the extent of the reduction in the price of the technology over time.

While Götz's model does account for some product differentiation, the model is not directly applicable to time-saving process innovations. For example, there is no explicit role in his model for a time-saving innovation nor does it account for differences in consumers' preferences for service quality related to their opportunity cost of time.[1] There is also no role for an innovation changing the degree of differentiation (i.e., changing the cross-price elasticities). In Götz's model increased competition is equivalent to less product differentiation, but the degree of product differentiation is not directly affected by the innovation itself. Since the technology is indivisible, there is also no direct implication for persistence in intra-firm adoption or through adoption of improvements in the technology.

More recently, Mulligan and Llinares (2003) establish another reason for adoption of a process innovation: horizontal differentiation in the services provided by the adopting firms. They find that innovations that increase service speed in the U.S. ski industry result in a different diffusion pattern than previously found in the literature. Mulligan and Llinares argued that innovations in the quality of service, such as service speed, might diffuse differently than innovations directed primarily at cost reduction. They report results concerning the diffusion of detachable chairlifts in the United States that provide the first empirical evidence that the adoption of a technological innovation by a firm decreases the likelihood that a local competitor will also adopt it. They model the effect that an innovation in service speed has on a ski resort's incentive to differentiate the quality of its service relative to that of its competitors and hypothesize that the incentive to adopt is negatively related to the number of competitors who have already adopted.[2]

As mentioned earlier, modification of the technology over time can complicate the use of duration models that implicitly assume that the underlying technology remains homogenous. Mulligan and Llinares avoid this problem by using chairlift technology, because there has been little change over time. The main difference has been in the cost of making the chairlifts, while the underlying characteristics have remained essentially the same. Their paper provides the starting point for the extended example of persistent adoption presented in Section 3.

Consumer willingness to pay higher prices in exchange for faster service is not new to the economics literature. For example, it is an important part of Davidson's (1988) duopoly queuing model with each firm offering service at different speeds and at different prices. The relative technological benefits of using service units of differing speed and number are also a part of earlier work by Mulligan [1983 and 1986]. While the change in speed creates a different quality of service that may appeal to all consumers, there is a limit to the number of consumers who value this higher quality of service enough either to pay relatively higher prices or incur greater travel costs.

Although modifications in technology over time create problems for duration analysis in general, there is also a problem that is peculiar to innovations that affect service time. In a working paper, Das and Mulligan (2004), we argued that the underlying technology could even change in significant enough ways to create a different diffusion pattern for each vintage of the innovation. Lumping vintages together can confound the analysis of the diffusion process. While cost-saving innovations may be apparent to consumers indirectly through lower prices, innovations in service quality are more directly obvious to them and may lead to more horizontal differentiation at least during the initial stages of the diffusion process.

We reported estimates from non-parametric and parametric duration models showing that the first vintages of optical scanners installed in supermarkets in the U.S. from 1974 to 1985 had a similar impact on service speed and followed a similar diffusion pattern to that of detachable chairlifts. Our results go beyond providing additional evidence of the product differentiation effect of time-saving technological innovations. In an earlier study, Levin, Levin, and Meisel (1987) considered a shorter diffusion time period and did not reach the same conclusions as we did. While we are able to reproduce their main results using their estimation procedure and comparable explanatory variables for their time period of analysis, the difference in our conclusions stems from our controlling for vintage effects.

The main features of the first vintages of optical scanner systems were faster processing of customer purchases and an itemized receipt. NCR released the first scanner system in June of 1974 with IBM providing a comparable system soon afterwards. The first scanning system technology, however, was prone to errors. The scanner would often not scan an item even after three tries requiring the checkout clerk to key in the price manually. Vintage 1 also had only one laser beam to read bar codes.

The first significant improvement (vintage 2) appeared in November 1979, when IBM released Model 3667, which increased service speed of the checkout process by allowing the scanner to read much smaller bar codes.[3] Vintage 2 had more laser beams and more accurate readings. The second vintage, however, diffused only between November 1979 and November 1980 before being replaced by the third vintage (IBM Model 3687), which diffused beyond the end of our sample period in March 1985. With vintage 3 the proportion of inaccurate readings decreased significantly. Vintage 3 was the first to use a holographic technique to reduce the handling of the items by the checkout clerks (Fistell, 35). It could also read bar codes that were damaged and read bar codes on items such as frozen meat (even if the bar code became crinkled and wet). All of these improvements led to faster processing time.[4]

The main point of this paper was that accounting for the unique characteristics of different vintages of an innovation is important, because these vintages may follow different diffusion patterns. While the third vintage increased the scanner's capacity to read ever smaller bar codes resulting in decreased processing time, by this time firms were able to take advantage of the cost-saving potential associated with hiring fewer workers needed to mark prices on individual items and with improvements in software used by the store processor. While processing speed did improve with subsequent vintages, the marginal improvement in service speed decreased significantly as the pace of improvements in processor capacity and software accelerated.

As noted earlier, Hannan and McDowell and Karshenas and Stoneman (1993) attributed the positive effect of prior adoptions on the rate of diffusion to an epidemic flow of information to potential adopters. As shown by Götz, a decrease in the price of a purely cost-reducing innovation or an increase in the stock of adopters increases the adoption rate for reasons that are unrelated to an epidemic spread of information.[5] Given that the main effect of the initial vintages of optical scanner technology was on service speed, we hypothesized and found that adoptions by competitors lowered the adoption rate in the same way reported by Mulligan and Llinares. As cost reduction became more important with subsequent vintages, this effect diminished.

Differences across vintages were not limited to prior adoptions by competitors. Given the effect of optical scanners on consumer time costs and an assumed relationship between income and the opportunity cost of time, we expected and found that the first vintage diffused more quickly in markets with higher household income. On the other hand, we expected that the coefficient for household effective buyer income becomes less important or even changes sign for later vintages that lowered costs, given relatively lower-income consumers' sensitivity to price. Anecdotal evidence at the time suggested that scanners were first installed in stores in relatively wealthier urban areas with only 29% of all stores having scanners by 1985. We also included a variable not normally found in the literature as another control for the opportunity cost of time: average household size. Having controlled for per-capita income, we considered larger household size as a proxy for the time costs of the person doing the shopping. As in the case of per-capita income, we expected and found this variable to be positively correlated with the diffusion of the time-saving first vintage but less so with subsequent vintages.

Götz's model suggests a faster diffusion for a cost-reducing innovation in competitive markets where increased competition is defined as less product differentiation. According to his model, greater competition forces firms to lower costs on the margin in order to remain competitive. On the other hand, there is ambiguity in the empirical literature about the impact of other measures of competition, such as the four-firm seller concentration ratio.[6] In our study we used a measure of the four-firm seller concentration ratio as our indicator of the degree of competition. While measures such as the four-firm seller concentration ratio are not exactly what is implied by Götz's measure of competition, we expected and found that more competitive markets are likely to diffuse the third vintage faster than the first one due to the pressure that competition has on pricing and the impact of the innovation on the firm's costs. On the other hand, Mulligan and Llinares argued that the number of competitors captures another aspect of the product differentiation effect, since a firm may adopt an innovation to differentiate the quality of its service before its rivals adopt the innovation. While the four-firm seller concentration ratio is to some extent negatively correlated with the number of competitors, Mulligan and Llinares argue that the more competitors a company has the more room it has for differentiating its service from that of competitors. We expected and found this variable to be positively related to the adoption probability.

While the theoretical literature focuses on establishing the possibility of a diffusion pattern for identical firms, Götz does incorporate a firm size variable in his model and shows that larger firms diffuse the innovation more rapidly as long as the technology is indivisible (that is, independent of firm size). The empirical literature hypothesizes the same relationship for essentially the same reason attributing it to scale economies. Even though anecdotal evidence suggests a potential scale economy at the store level given the need for a central processor for at least the third vintage, we did not have a direct measure of store size. While we did have data for the average store size per SMSA, we do not have this information on a per-store or per-company basis.

Most empirical studies in this literature find that the incentive to adopt the innovation increases with other less direct measures of firm size, such as firm market share.[7] In our case, we included both firm market share and a second possible control for firm size: whether or not the company is a large supermarket chain. While anecdotal evidence suggests that scale economies start to become more important for companies at the SMSA level after 1980 and especially after 1985 due to increased capacity for the store processor and software development linking scanner data from a chain's stores in a specific SMSA, the full potential at the SMSA level was still in the earliest stages of development by the beginning of 1985. On the other hand, since greater access to financing for a new and unproven technology may be a factor in explaining adoption of the technology, we hypothesized that these variables may have been factors in explaining the diffusion of at least the first vintage of this technology.

Anecdotal evidence also suggests that some stores may have reduced labor costs by reducing the number of cashiers.[8] While we did not have adequate data on labor costs or the number of cashiers per store for the entire time period of the study, we did have data on the average number of square feet per number of checkouts in the SMSA at the start of each time period under investigation. Since stores with more square feet per checkout may be economizing on relatively higher labor costs, we expected and found this variable to be positively correlated with the rate of diffusion.

Saving costs by eliminating the need for marking prices on packages was also an initial expectation in the industry, but consumer resistance delayed the realization of these benefits. Anecdotal evidence suggests that this resistance did not persist throughout the time period. According to the March 27, 1978 article,

"Just 208 stores, considerably less than 1 percent of the country's 33,000 supermarkets, have so far been equipped with registers that can read the [Universal Product] code. Now, however, several years behind schedule this revolution in grocery retailing is gathering momentum. The Retail Clerks International Union, which had feared that scanners would be bad for employment, has reconsidered the evidence on that point and retreated to a posture of neutrality. Labor's allies in the consumer movement have also quieted down—perhaps because the consumers they presume to speak for have fallen in love with scanners" (Coyle, 76).

Despite growing acceptance of scanners, six states (California, Michigan, New York, Connecticut, Massachusetts, and Rhode Island) passed laws by

1976, still in effect, requiring that supermarkets with scanners place prices on each item as a consumer protection initiative. To test the impact of this law we included a dichotomous variable for the six states that passed laws requiring stores to mark prices on individual items. We find that these laws had no effect on the adoption of scanners during the diffusion of the first vintage, since consumer resistance in all states during this time period made such laws redundant. However, we expected and found that these state laws had a negative impact on the diffusion rate of the third vintage.

We also accounted for possible persistent adoption by estimating the model with two additional variables controlling for adoption decisions concerning earlier vintages of the technology both for the company and for its competitors. Including these variables does not change the sign or the level of statistical significance of the coefficients of any of the other covariates. While the incentive to adopt decreases with prior adoptions of vintage 3 by rivals, it increases with the number of rival companies that adopted vintage 1 or vintage 2 and whether the company itself had adopted one of the earlier two vintages. These results are consistent with persistence of the incentive to differentiate a store's quality of service relative to that of its competitors.

While we find that none of the covariates affecting the adoption decisions of these vintages lose their predictive power when control variables are added for adoptions of other vintages, decisions concerning the adoption of later vintages are affected by the firm's and its rivals' decisions concerning earlier vintages. In particular, we find no evidence that supermarket chains waited for the second or third vintage of the technology before making their first adoptions. Firms adopting the first vintage of the technology were more likely to adopt the subsequent vintages, and symmetrically those firms adopting the later vintages were more likely to have adopted an earlier vintage. Interestingly, firms adopting later vintages were less likely to do so if their competitors had adopted the later vintage, but were more likely to adopt the later vintage if their competitors had adopted an earlier vintage. We suggest that these results are consistent with our overall hypothesis that firms may persist in adopting subsequent vintages of time-saving process innovations if it allows them at least for some time to differentiate their service from that of their competitors.

After March 1985 information on adoptions of scanning technology was no longer provided by the trade publications even though less than a third of the approximately 30,000 stores in the United States had scanners at that time. While we are unable to analyze formally what happened after March 1985, anecdotal evidence suggests that the pace of adoption increased dramatically primarily due to rapid expansion in store processor capacity and development of more advanced software. For example, by July 1994 25,000 supermarkets were using scanners at that time representing 95 percent of chain stores and 75 percent of independent supermarkets. After 1985, marginal improvements in scanning speed were minimal compared to purely cost-saving improvements.

As this extended example illustrates, innovations can have effects that vary considerably from one vintage to the next. While not all innovations are subject to this problem, this example suggests that detailed knowledge of the effect that an innovation has on the production process is an important starting point for an empirical study of the diffusion process. Studies that rely on aggregated data that may mask the impact of specific innovations are thus subject to questions about their applicability. We now turn to intra-firm diffusion, where the aggregation problem can be especially serious.

2.6 INTRA-FIRM DIFFUSION.

In their recent paper Battisti and Stoneman state that "the study of intra-firm diffusion has largely been neglected and the limited extant literature overwhelmingly relies upon uncertainty reduction via information spreading or epidemic learning as the main driver." For example, this literature gives particular importance to a variable, originally proposed by Mansfield (1963), measuring the time since initial adoption of the technology as a proxy of information flow within the company. Despite the common finding in this literature that the extent of use of a specific technology increases with the number of years since first adoption, Battisti and Stoneman find no evidence of this with their data.

Battisti and Stoneman adapt the approach taken by Karshenas and Stoneman (1993) for inter-firm diffusion to a study of intra-firm adoptions of computer numerically controlled technology (CNC) at firms in the UK engineering and metalworking sectors. Battisti and Stoneman base their approach on the assumption that firms will gain differently from using more of a specific technology due to differences across firms. The value of additional amounts of the technology will also depend on the level or the stock of the technology already used by the firm. Given the cost of the technology, not all firms will adopt the same level of the technology at a point in time. Battisti and Stoneman apply this logic to an equilibrium model with firms deciding whether or not to add to their capital stock over time.

In their empirical work the dependent variable is the log of the ratio of the machine-tool stock of the firm incorporating CNC technology in 1993 to that not using CNC technology. The independent variables include measures of firm size (employees), current and expected user costs, years in business, use of in-house R and D, use of complementary technologies, and two proxies for epidemic effects: log years from the firm's first adoption of CNC technology up to 1993 and log of the within-industry share of adopters at the time of the firm's first adoption. Expectations of future price changes are based on adaptive expectations using historical changes in prices. The industry is defined as all firms producing the majority of their products in the same three-digit (and in some cases, four-digit) SIC code industry. As generally found in the inter-firm diffusion literature, firm size is correlated with adoption. They find no support for an epidemic effect due to information spreading or learning effects as measured by their two proxies. In their case, the percentage of firms adopting in the same three-digit SIC industry and the time since first adoption do not have an effect on the firm's own extent of adoption.

As was the case for Bartoloni and Baussola, aggregation poses a problem for Battisti and Stoneman. Since the CNC technology first appeared in 1970, it should have undergone several changes by 1993. Although acknowledging that changes in the technology could affect the diffusion pattern, they do not have information about the various vintages of the technology or how the different firms in their sample use the technology. Battisti and Stoneman implicitly assume that firms within the three-digit industries are essentially producing the same kinds of products and are direct competitors. However, in their sample only one of 343 firms had 100 percent usage by 1993, while 59 were excluded from their initial sample for considering the technology inappropriate for their production process. By 1993 the proportion of the machine tool stock of the plant using CNC was less than 20 percent for slightly more than 50 percent of the firms with only 7 percent with a proportion over 70 percent. It is not clear from their paper whether or not these proportions differ significantly by SIC code or whether adopting firms produce some of the same products as the 59 firms that do not use CNC technology at all. As a result, we do not know if all the firms would ever adopt the technology at a 100 percent rate or if they had already reached their theoretical upper limit by 1993. In other words, it is not clear if the adoption rate is actually a proxy for the mix of products produced in these three-digit SIC industries.

Earlier papers on intra-firm diffusion that relied on less aggregated data are subject to a similar criticism. For example, Levin, Levin, and Meisel (1992) estimated a model of intra-firm diffusion using data from their previous study Levin, Levin, and Meisel (1987) on inter-firm diffusion of point-of-sale optical scanners. Following Mansfield, they attributed the intra-firm diffusion pattern to an epidemic learning effect within the firm. As shown in Das and Mulligan (2004), Levin, Levin, and Meisel (1987) estimated their inter-firm duration model implicitly assuming that the technology was homogeneous even though their time period includes the diffusion of three different vintages. Levin, Levin, and Meisel (1987) treated each adoption by the firm as an additional adoption of the original technology, while some of these adoptions were actually upgrades or replacements. Since they used the same data, Levin, Levin, and Meisel (1992) did not know the ratio of a company's stores that had actually converted to optical scanners at any point in time.

2.5   SUMMARY

The literature reviewed to this point contains a wide range of theoretical models and empirical examples showing how firms may have an incentive to persist in the adoption of new technologies given assumptions about the learning process associated with specific technologies and across vintages, expectations about future prices and the extent and timing of improvements in quality, and adoption decisions by rivals. We emphasized the importance of knowing how the innovation actually affects the firm's production process. While there is a theoretical literature that considers the diffusion pattern of specific innovations, most empirical studies rely on aggregate data that can mask the impact of the innovation on the firm's production process. This problem can be especially serious for time-saving innovations, since the nature of the diffusion process may change as the technology evolves over time. Our earlier work on optical scanners is a case in point.

In the next section we turn to another innovation that increases service speed: the high-speed chairlift technology now widely used in ski resorts. Unlike optical scanners and most of the innovations studied in the literature, chairlift technology has not undergone continuous improvement. Improvements in the technology have been limited to an increase the number of seats per chairlift from four to six. As a result, persistence in this case takes the form of an increasing percentage of chairs using the technology at any one ski resort. Unlike the main focus of papers on intra-firm diffusion of technology, such as scanners or ATMs where diffusion means converting a firm's additional stores or banks to the technology, adding another high-speed chairlift affects everyone at the ski resort. In this sense intra-firm diffusion results in an increase in quality of service as the number of chairlifts increases at the ski resort. The impact of intra-firm diffusion on vertical quality differentiation plays an important role in our description of the nature of competition in this industry.

3. HIGH-SPEED DETACHABLE CHAIRLIFTS

In this section we consider persistent adoption of high-speed chairlift technology in the U.S. ski industry. We expand on earlier work by Mulligan and Llinares concerning the inter-firm diffusion of this technology from 1981 to 1997. In Section 3.1, we summarize Mulligan and Llinares' main results. As mentioned earlier, they reported estimates of a parametric diffusion model consistent with the hypothesis that the incentive to adopt high-speed detachable chairlifts decreases as other local ski resorts adopt. By contrast, the empirical literature generally finds that regardless of the nature of the innovation, the rate of adoption is an increasing function of prior adoptions by local competitors. Our main focus, however, is the likelihood of persistent adoption of high-speed technology after the initial adoption. In Section 3.2 we provide empirical support for persistent adoption at ski resorts that cater primarily to avid and vacation skiers.

3.1 FIRST ADOPTIONS

In an earlier model of lift ticket pricing with homogeneous chairlift technology, Barro and Romer (1987) made a distinction between avid and other skiers based on a skier's willingness to pay for additional runs per day. They hypothesized that some ski resorts would specialize in avid skiers by charging higher prices leading to less congestion and more runs per skier per day.

Mulligan and Llinares argue that the high-speed detachable chairlift technology provides an additional means of creating more runs per day for avid and vacation skiers. The detachable chairlift offers skiers features not provided by the older fixed-grip technology, such as easier loading, a potentially more exciting ride, somewhat fewer loading and unloading incidents, and somewhat higher actual (as opposed to design) capacity due to lower frequency of breakdowns, while reducing the time needed to go to the top of the hill by approximately one-half.[9] While these characteristics may appeal to many skiers, Mulligan and Llinares reported that only 68 of over 400 U.S. ski resorts had installed a detachable chairlift by 1997. By 2005 this number had increased to 99. A major deterrent to adoption is the cost, since the detachable chairlift has remained approximately 40 percent more expensive per unit of capacity relative to the older technology.[10]

Because of the cost increase, models focusing on cost-reducing process innovations are unable to explain the diffusion of this type of technology. As additional ski resorts adopt detachable chairlifts to attract skiers willing to pay higher prices for these chairlifts' special features and the additional runs per day, late adopters must share this submarket with those that have already adopted. With a continuum of skier types the adopting ski resorts would have to lower their prices in order to attract relatively less avid skiers as the number of adopters increases. Given the relatively smaller number of skiers placing a high value on lift speed in local markets and the relatively high cost of the chairlift, the incentive to adopt is likely to decrease as other local ski resorts adopt.

To test this hypothesis Mulligan and Llinares estimated a parametric duration model using annual data for 344 ski resorts having at least one chairlift from the 1980-1981 to 1996-1997 ski seasons for the following exogenous variables: whether or not the ski resort started business after 1970, whether or not the ski resort is on U.S. Forest Land, vertical drop, the number of potentially skiable acres, the number of competitors within 125 miles of the ski resort, and number of adopters within 125 miles of the ski resort at time t-1.

Given that new ski resorts have generally added chairlifts in stages over time, they conjectured and found that the newest ski resorts would be more likely to add chairlifts of any type in a given year. They used a dichotomous variable indicating whether or not the ski resort started business after 1970 and also re-estimated the model with similar results using the actual establishment date. The median initial date of operation was 1958.

Ski resorts on National Forest land generally have lower costs, because the rental prices for use of the land have not reflected market prices. On the other hand, they face constraints concerning parking and hotel rooms that other ski resorts do not face, because of their remote locations and restrictions imposed by the U.S. Forest Service. Specializing in a smaller number of avid skiers may result in less congestion of these related facilities and provide them an additional incentive to adopt the detachable chairlift. While only 32.8 percent of ski resorts in the sample were located on land operated by the U.S. Forest Service, Mulligan and Llinares found that these ski resorts were more likely to make initial adoptions, ceteris paribus.

Mulligan and Llinares used vertical drop as a control for both the size of the ski resort and the expected increase in demand during this time period. Ski resorts with higher vertical drops may also have an incentive to adopt the detachable chairlift due to scale economies.[11] Although the number of skier days has remained relatively constant in the U.S. throughout this time period, skiers are skiing more during extended vacation periods at ski resorts with larger vertical drops. These are also the ski resorts most likely to be competing with the largest ski resorts in the U.S. and the rest of the world for the avid skier. As a result, ski resorts with larger vertical drops were more likely to make an initial adoption. In the next section we find the same result for the proportion of high-speed chairlifts.

While ski resorts have some flexibility in increasing their lift capacity and skiable acres at the beginning of the ski season, they will eventually face a binding limit on skiable terrain. At higher lift capacity to potentially skiable acre ratios, skiers face a relatively greater degree of congestion that limits the number of ski runs. Since lift capacity in any given year is endogenous, they did not include lift capacity in the estimation. Mulligan and Llinares hypothesized and found that ski resorts with more potentially skiable acres will be more likely to add a chairlift regardless of its speed. Since there is no direct exogenous measure of potentially skiable acres available, they used actual skiable acres in the year 2000 as their proxy for this variable and terminated the sample period at the end of the 1996-1997 ski season. In the next section we find that skiable acreage is also an important factor limiting the persistence of high-speed chairlifts at all but the largest ski resorts.

Due to a lack of data on skier visits and skier types at specific ski resorts, Mulligan and Llinares were unable to determine the number of skiers or the distribution of skier types per local market. They argued, however, that ski resorts with a larger number of competitors charge higher lift ticket prices, ceteris paribus, likely due to higher demand per ski resort. Since ski resorts catering to avid skiers in the Barro-Romer model charge higher prices even in the absence of faster chairlifts, the higher prices could also indicate a higher concentration of relatively avid skiers in these local markets. As a result, Mulligan and Llinares hypothesized that an increased number of competitors, ceteris paribus, provides a greater incentive for any ski resort to adopt the faster chairlift in order to differentiate its service relative to that of its competitors. They defined this variable as the number of ski resorts located within 125 miles of driving distance from the ski resort and also estimated the model with local competitors defined as all ski resorts within 50 miles of the ski resort.[12] They defined the number of adopters as all ski resorts within 125 miles of the ski resort that had installed at least one high-speed detachable chairlift by the end of the previous ski season. They also estimated the model using 50 miles in place of 125 miles.

Mulligan and Llinares also found that the hazard function exhibited the property of positive duration dependence generally found in the literature. Karshenas and Stoneman (1993) argued that by allowing the hazard rate to vary over time, one provides a better test of the effect of prior adoptions by competitors than that of earlier studies that had restricted the distribution of the hazard function. While they also found positive duration dependence for some of the industries in their sample, there was an absence of a deterrent effect of the number of prior adoptions on a firm's adoption rate. Karshenas and Stoneman justify this finding by making an analogy to the spread of epidemics based on "(i) the learning processes involved in the use of new technology and its transmission through human contact, with the "infection" being information; (ii) pressure of social emulation and competition; or (iii) reductions in uncertainty resulting from extensions of use" (p. 509).

Mulligan and Llinares believed that positive duration dependence for the ski industry is due to a different reason. When the detachable chairlift was introduced in 1981, there were several articles about the new technology in the trade literature, and the technology had already been in use at European ski resorts. As a result, it is unlikely that a lack of information alone, at least among ski resorts, could explain the positive duration dependence. There is evidence to suggest that changes in skier opportunity costs of time during the 1980s and 1990s encouraged ski resorts to increase overall capacity in order to shorten lift lines and increase the potential number of runs per skier per day. For example, although the number of skier days at U.S. ski resorts remained relatively constant at approximately 50 million per year between 1980 and 1997, overall lift capacity in the industry increased by approximately 70 percent. Lift ticket prices increased by 154.69 percent on non-holiday weekdays and 148.86 percent on holidays and weekends during this time period, while the CPI increased by only 78.4 percent. In addition, the nominal price of both fixed-grip and high-speed detachable chairlifts remained essentially unchanged during this time period.

Willingness of skiers to pay even higher lift ticket prices, ceteris paribus, at ski resorts with detachable chairlifts, especially during off-peak time periods, suggests that increases in the opportunity costs over time have resulted in an increased demand for more runs per day. While only 5 percent of the U.S. population goes skiing at least once each year, skiers are among the wealthiest Americans. In 1996 average skier household income was more than $80,000 with the share of skier households with income under $50,000 only 27 percent (Cravatta, 1997). By comparison, median U.S. household income was only $35,000 in 1996. Since 1980, the biggest increases in income were received by the upper fifth of the income distribution.[13] Given no change in skier days per year, the increase in overall capacity has increased the potential number of runs per skier-day for all skiers. Since a detachable chairlift can result in even more runs per day with the same capacity as that of the older technology, the empirical results are consistent with the overall increased demand for more runs per day in the industry as a whole.

Since Mulligan and Llinares' main focus was the effect of local competitors on the adoption decision, the definition of the local market was a concern. For example, the distances between clusters of Western and Rocky Mountain ski resorts are more pronounced than in other regions. When the market area includes all ski resorts within 125 miles of one another, there is little overlap of local markets in the Western and Rocky Mountain regions.[14] This is not the case elsewhere. For example, all ski resorts in New Hampshire share the same competitors within New Hampshire, yet those located in the Northeastern part of the state have competitors in Maine within 125 miles, while those in the Southwestern part have local competitors in Vermont and

Massachusetts. Despite these differences, Mulligan and Llinares report results showing that the coefficients for the number of local competitors and number of local adopters variables are statistically significant with the same predicted signs for the Eastern, Western and Rocky Mountain regions.

Another potential problem in the estimation of diffusion models is the case of simultaneous adoptions. Mulligan and Llinares assumed that adoptions are made without knowledge of the intentions of one's competitors, which is also Hannan and McDowell's and Levin, Levin, and Meisel's maintained assumption. By contrast, Karshenas and Stoneman (1993) assumed that firms knew whether or not their competitors would also adopt at time t and t + 1. This approach, however, amounts to assuming perfect foresight and ignores the strategic interaction among competitors. Karshenas and Stoneman justify this approach by stating that while it "does not fully reflect the strategic nature of some of the recent [theoretical] contributions, ... , the theoretical literature gives no indication of what might be a more adequate empirical approach" (p. 507).

Simultaneous adoptions were a potential problem given that 28 of the 68 ski resorts in the sample adopted in the same year as at least one other local competitor. Given that some of the ski resorts may have known of the binding intention by local competitors to adopt, Mulligan and Llinares also estimated the model based on the extreme assumption that every ski resort knew whether or not its competitors were going to install a detachable chairlift prior to making its own decision. The coefficient for prior adopters remains positive and statistically significant for the Eastern ski resorts at the five percent level and for the entire sample at the ten percent level.

While the coefficients for the number of prior adopters are still positive for the Rocky Mountain and Western regions, they are no longer statistically significant. One possible interpretation is that these ski resorts are less concerned about the adoption decisions of their local competitors and are only concerned about the national and international market. However, even if some of the ski resorts knew in advance of binding commitments by competitors to add a faster ski lift, not all of the ski resorts could have had this information. Even assuming that everyone knew everyone else's intentions, there was still no support for the alternative hypothesis of an epidemic effect due to the adoption decisions of one's most immediate competitors. In the next subsection we present evidence to suggest that these ski resorts do follow a different pattern of persistent adoption compared to those with a primarily local clientele. However, we also reject the possibility of an epidemic informational effect.

Mulligan and Llinares report that a large number of small ski resorts left the industry between 1980 and 1997, while 26 ski resorts entered. Six of the entrants eventually installed a detachable chairlift. While duration models control for right-hand censoring associated with some non-adopters still in the sample at the end of the sample period, there is no direct method for controlling for left-hand censoring. Mulligan and Llinares reported results including these 26 ski resorts, because all of the exogenous variables are either site specific (vertical drop, location on National Forest land, and potentially skiable acres) or competitor specific (number of competitors and number of prior adopters). None of the entrants rented U.S. Forest land. Given that someone could have used these lands as ski resorts at any point during this time period, the decision to add a detachable chairlift is essentially the same as that faced by an existing ski resort.   They also estimated the model without these 26 ski resorts with no significant change in the estimated coefficients.No ski resort with a vertical drop below 500 feet or fewer than 75 skiable acres installed a detachable chairlift during this time period.   As a further test, they estimated the model excluding the 122 ski resorts with less than 500 vertical feet and 75 skiable acres with essentially identical results for the number of prior adopters and local competitors.

3.2 PERSISTENT ADOPTION

Since Mulligan and Llinares did not consider the persistence of adoption either within firms or across vintages of high-speed technology, we now investigate the extent to which individual ski resorts have altered the mix of chairlifts using high-speed technology since 1981. While the technology was first used for chairlifts with four seats (HS quads), more recently manufacturers have added high-speed chairlifts with six seats (HS six-packs). Although there is a wide array of technology in use from rope tows and J-bars to gondolas and trams even at the same ski resort, we concentrate in this section on the proportion of chairs using high-speed technology. This comparison is likely to understate the extent of use of the technology given that we do not control for the vertical drop covered by each type of technology. While this information is available for some ski resorts, it is not generally provided.   We assume that surface lifts are designed for short vertical drops and beginners and are thus not likely to be converted to high-speed technology.

Table 1 shows the number of ski resorts by state that have adopted at least one high-speed chairlift by 2005. As reported by Mulligan and Llinares, 27 of 72 ski resorts in the Rocky Mountain states and 17 of the 58 Western ski resorts had adopted by 1997, compared to only 21 of the 115 Eastern ski resorts and 3 of the 87 ski resorts in the North Central region. By 2005 the number of initial adopters increased to 32 in the East (of which 20 were in the three Northern New England states), 9 in the North Central region, 33 in the Rocky Mountain region, and 23 in the Western region.

Table 1. Ski areas with high-speed chairlifts by state

STATE	NUMBER OF SKI RESORTS
Alaska	1
Washington	4
Oregon	4
California	12
Nevada	1
Arizona	1
Total Western Region	23
Montana	3
Utah	8
Wyoming	2
Colorado	15
New Mexico	1
Idaho	4
Total Rocky Mountain Region	33
Maine	2
New Hampshire	9
Vermont	10
Total Northern New England	21
Massachusetts	2
New York	5
Pennsylvania	3
New Jersey	1
Virginia	1
West Virginia	1
Total Eastern Region (excluding No. NE)	13
Michigan	4
Minnesota	1
South Dakota	1
Wisconsin	3
Total North Central Region	9
Total	99

One possible form of persistent adoption concerns the types of chairlifts used at a ski resort. The main innovation in chairlift technology since 1981 is an increase in the number of seats from four (HS quad) to six (HS six-pack). Table 2 provides a cross-tabulation of ski resorts based on the numbers of high-speed quads and six-packs. 234 of the 333 ski resorts having at least one chairlift in 2005 did not use high-speed technology. Eight of the nineteen ski resorts with HS 6-packs did not have any HS quads. In addition, nineteen of the thirty HS 6-packs were located at ski resorts with at most one HS quad, while the number of HS quads per ski resorts ranged to a maximum of fourteen. We interpret this information to mean that the adoption of the HS 6-pack technology was unlikely to represent persistence in adoption of new technology distinctly different from that of HS quads. In other words, given the large percentage of first-time adopters of high-speed technology choosing HS 6-packs, it appears that prior familiarity with HS quads is not motivating the adoption decision.

Table 2: Cross-tabulation of high-speed quads and six-pack. *High-speed six-packs

High-speed Quads	0	1	2	3	4	Total
0	234	6	1	0	1	242
1	32	3	0	0	1	36
2	20	0	1	0	0	21
3	7	0	1	0	0	7
4	9	1	0	1	0	11
5	1	2	0	0	0	3
6	2	0	1	0	0	3
7	5	0	0	0	0	5
8	1	1	0	0	0	2
9	2	0	0	0	0	2
14	1	0	0	0	0	1
Total	314	13	3	1	2	333

Quads and Six-Packs refer to the capacity of each of the chairs on a ski lift, either 4 or 6 people.

As a result, we measure persistence as the proportion of chairs using high-speed technology regardless of the mix between HS quads and six-packs. Our approach has similarities to that of Battisti and Stoneman who used the ratio of machines with CNC technology to those without it. In our case we use the proportion of chairs that incorporate the high-speed technology. We measure this variable in two ways. First, we use a simple proportion of chairs. Second, we double the number of chairs both in the numerator and denominator using high-speed technology. The motivation for this adjustment comes from the earlier discussion concerning the capacity equivalence of high-speed and fixed-grip technology. The HS detachable chairlift moves at approximately twice the speed of the fixed-grip technology. As a result, a HS quad has the equivalent capacity of two fixed-grip quads in terms of number of skiers transported per time period. This second is, therefore, a more accurate measure of the proportion of chairlift capacity using high-speed technology. In the following tables we report only those results for the second measure of persistence. In our regression analysis, we find that the signs and statistical significance of the coefficients are the same in all cases.

Table 3 shows the number of ski resorts and minimum, maximum, mean and standard deviation for the second measure of persistence for the 99 ski resorts with at least one high-speed chairlift in 2005. The table also shows this information for each of five regions. For example, while the mean value for the Rocky Mountain ski resorts is 63 percent, the national average is 53 percent. Ski resorts in the Western States (Alaska, Washington, Oregon, California, Arizona, and Nevada) have a mean value of 56 percent. The North-central ski resorts, on the other hand, have a mean value of only 39 percent.

Table 3: Proportion of high-speed chairlifts by region

Region	Number	Minimum	Maximum	Mean	Standard Deviation
Western	23	0.23	0.86	0.56	0.19
RockyMountain	33	0.31	1.00	0.63	0.17
Northern New England	21	0.33	0.78	0.49	0.12
Eastern (minus No. NE)	13	0.22	0.73	0.41	0.14
North Central	9	0.25	0.67	0.39	0.14
All Regions	99	0.22	1.00	0.53	0.18

Table 4 shows persistence as the sum of chairlifts multiplied by the number of seats per ski resort for each of six regions of the country. Of the 99 ski resorts, 31 have only one HS quad, while 7 have only a HS 6-pack. The remaining 61 ski resorts have more than one high-speed chairlift. While 22 ski resorts in the North-Central and Eastern (excluding Northern New England) Regions had adopted at least one high-speed chairlift by 2005, only 5 had two and one had three. In the Northern New England Region 11 of 20 ski resorts had more than one high-speed chairlift, while 27 of 33 ski resorts in the Rocky Mountain region and 17 of 23 in the Western Region had more than one high-speed chairlift.

Table 4: Number of chair-weighted lifts per ski resort by region

High-Speed Chair	Western	Rocky Mountain	Northern New England	Eastern	North Central	Total
4	5	5	9	7	5	31
6	1	1	1	2	2	7
8	4	8	3	3	2	20
10	1	2	0	0	0	3
12	2	4	1	1	0	8
16	4	1	4	0	0	9
20	1	0	1	0	0	2
22	0	1	0	0	0	1
24	1	0	2	0	0	3
26	1	1	0	0	0	2
28	1	5	0	0	0	6
32	1	0	0	0	0	1
34	1	0	0	0	0	1
36	3	0	0	0	0	3
38	1	0	0	0	0	1
56	0	1	0	0	0	1
Total	23	33	21	13	9	99

In the literature review we noted that the number of years since first adoption has been used as a proxy for internal informational flows on the extent of intra-firm diffusion of an innovation. For example, this variable was used by Battisti and Stoneman in their study of CNC technology. Following Mansfield researchers have attributed a positive correlation between the extent of intra-firm diffusion and time since adoption as evidence of an epidemic effect of information. In Table 5 we present OLS regression results that also find a correlation between the proportion of HS chairs as the dependent variable and the year of first adoption. The other independent variables are vertical drop, average annual snowfall, number of skiable acres, and location on national forest land. In addition to year of adoption, which is statistically significant at the 5 percent level, vertical drop is the only other variable that is statistically significant.

While following Mulligan and Llinares one can make an argument for the exogeneity of the other variables, the year of initial adoption is endogenous and highly correlated with vertical drop. As Mulligan and Llinares report, vertical drop is a major determinant of the year of initial adoption. In addition, since it is not clear that ski resorts would need several years to learn about the usefulness of high-speed technology before adding more chairlifts, the usual argument of an epidemic effect given in the literature seems unwarranted. When a ski resort adds an additional high-speed chairlift, there is an increase in quality available to all the skiers not just to additional consumers unaffected by a firm's initial adoption of the innovation at a different location. Table 5 also provides results with year since adoption excluded. In this case vertical drop increases in significance, while no other regressor has a statistically significant coefficient.

Table 5: Ols regression with and without year of first adoption (dependent variable: proportion of high-speed chairlifts)

Variable	With Year of First Adoption	Without Year of First Adoption
	Coefficient (standard error) (t-statistic)	Coefficient (standard error) (t-statistic)
Constant	17.526** (8.215) (2.133)	0.379*** (0.048) 7.936
US Forest Land (Yes = 1)	0.028 (0.039) (0.726)	0.041 (0.038) (1.086)
Vertical Drop (Feet in Thousands)	0.048* (0.027) (1.755)	0.061** (0.025) (2.420)
Annual Average Snowfall (Hundreds of Inches)	-0.006 (0.015) (-0.420)	-0.010 (0.015) (-0.647
Skiable Acres (Thousands)	0.021 (0.021) (1.004)	0.025 (0.021) 1.216
Year of Initial Adoption	-0.009** (0.004) (-2.107)	
Adjusted R-Square	0.259	0.259

Mulligan and Llinares argued that vertical drop is a proxy for the overall quality of the services provided by the ski resort, increases in demand over time and scale economies for high-speed chairlifts. Ski resorts that cater primarily to a local or regional submarket are less likely to have large vertical drops. In addition, differences in vertical drop for these ski resorts are unlikely to be large enough to explain persistence across local markets. To illustrate this point we make a distinction among ski resorts depending on their locations. Table 6 shows that for states located either in the Western Region (Alaska, Washington, Oregon, California, Arizona, and Nevada) or the Rocky Mountain Region (Utah, Montana, Idaho, Colorado, Wyoming and New Mexico) the only factor affecting the degree of persistence is vertical drop as was the case of the entire sample. However, the coefficient for vertical drop is not statistically significant for the 44 ski resorts in the remaining states where vertical drops are much smaller. Table 6 also includes regression results for all ski resorts except the fifteen located in Colorado that had adopted at least one high-speed chairlift by 2005. These ski resorts have the largest vertical drops and a high degree of persistent adoption. Interestingly, for the remaining 84 ski resorts skiable acres is now the dominant determining the extent of persistent adoption.

Table 6: Ols regressions (proportion of high-speed chairlifts as dependent variable)

Variable	Mean (Std. Dev.)	Western and Rocky Mountain Regions Coeffs. (Std. Errors) (t-stat.)	Mean (Std. Dev.)	North-Central and Eastern Regions Coeffs. (Std. Errors) (t-stat.)	Mean (Std. Dev.)	All States Except Colorado Coeffs. (Std. Errors) (t-stat.)
Constant		0.44*** (0.11) (3.965)		0.393*** (0.051) (7.744)		0.390*** (0.047) (8.235)
US Forest Land (Yes = 1)	0.821 (0.386)	0.05 (0.060) (0.088)	0.186 (0.394)	0.039 (0.055) (0.701)	0.476 (0.502)	0.013 (0.039) (0.321)
Vertical Drop (Feet in Thousands)	2.516 (0.859)	0.092** (0.035) (2.609)	1.501 (0.741)	0.006 (0.044) (0.137)	1.916 (0.885)	0.037 (0.029) (1.289)
Annual Average Snowfall (Hundreds of Inches)	3.522 (1.208)	-0.020 (0.020) (-0.996)	1.432 (0.702)	-0.007 (0.042) (-0.163)	2.571 (1.565)	-0.003 (0.015) (-0.201)
Skiable Acres (Thousands)	1.941 (1.184)	-0.005 (0.026) (-0.175)	0.310 (0.274)	0.152 (0.096) (1.592)	1.113 (1.169)	0.040* (0.024) (1.694)
Number	56		43		84	
Adjusted R-Square		0.121		0.017		0.151

*   ten percent level of significance
** five percent level of significance
*** one percent level of significance

factor Taken as a whole the econometric results suggest that a subset of U.S. ski resorts was particularly well suited to serve as a national market for avid and vacation skiers. Several factors, such as skiable acreage, annual snowfall and location on National Forest land, are highly correlated with vertical drop. All of these factors are site specific and indicators of quality. Nearly all of the ski resorts in the national submarket already existed when the first high-speed detachable chairlift was installed at Breckenridge, Colorado in 1981. As a result, ski resorts located in the Rocky Mountain and Western regions were especially well situated to benefit from the increase in demand for skiing due to falling real transportation costs and higher real incomes. Given the relatively higher vertical drops in the Western and Rocky Mountain regions, the greatest extent on persistent adoption occurs in these areas.

4. CONCLUSION

In this chapter we considered an extensive literature related to the adoption of process innovations. Despite a wide range of theoretical models with different assumptions and conclusions, empirical diffusion studies of specific innovations have been limited for two main reasons. Data are often aggregated across firms and innovations without providing sufficient detail of the nature of the impact of the innovation on the firm's production process. A second related problem concerns the possibility that the innovation could undergo constant modification during the diffusion process making it difficult to distinguish between diffusion of the basic technology and persistent adoption of improvements. Both problems can be especially important for an empirical study of persistent adoption.

We illustrated the second problem with a detailed example from an earlier working paper showing how the nature of the diffusion process for point-of-sale optical scanners can change substantially from vintage to vintage. By lumping these different vintages together, one provides a misleading interpretation of the diffusion process and the impact of important covariates, such as the number of prior adopters, consumer income and family size, and the degree of competition.

We also provided new evidence for the persistent adoption of high-speed chairlift technology that avoids both problems. Unlike optical scanners and most other innovations, chairlift technology has not undergone significant change since its introduction in the U.S. in 1981. The main improvement has been in the cost of making the lifts with competition keeping the nominal price relatively unchanged during this time period. The performance characteristics of high-speed chairlift technology were also well known to the ski industry at the time of the first adoption mitigating concerns voiced in the evolutionary diffusion literature that potential adopters are not sufficiently well informed or capable of optimizing in a manner assumed by neoclassical diffusion models. Despite twenty-four years of diffusion within the U.S. ski industry, only 99 of over 300 ski resorts with at least one chairlift have adopted this technology by 2005. A study by Mulligan and Llinares has shown that initial adoptions of this technology slowed down the rate of initial adoption by local competitors as early adopters attracted a subgroup of relatively avid skiers willing to pay higher lift-ticket prices.

The ski industry also has a somewhat unique structure that it shares with other recreational and entertainment industries due to the presence of both local/regional and national/international submarkets. We argue in this chapter that these two markets are not likely to provide the same level of service quality due to greater heterogeneity in the mix of skiers at local ski resorts. Although Mulligan and Llinares found that ski resorts in regions that cater primarily to vacation skiers made initial adoptions in a manner similar to that elsewhere, we expect to see and find a higher proportion of high-speed technology at national ski resorts persisting in their adoption of the technology over time. The national market is likely to attract vacation and avid skiers willing to pay for a higher quality skiing experience. While ski resort quality has several dimensions, we have given prominence to vertical drop and skiable acreage as proxies for quality that distinguishes national ski resorts from local ones.

Our empirical results also relate to an on-going discussion in the literature concerning the importance of epidemic informational effects on intra-firm diffusion of technology. Most papers in this literature find that the diffusion rate is an increasing function of the time since the initial adoption of the technology. The generally accepted interpretation of this result is a flow of information within the firm concerning the merits of the innovation. We find the same empirical result following an estimation procedure similar to that used in the literature. On the other hand, we interpret this result quite differently. While ski resorts that made their initial adoptions earlier also had a higher proportion of their lifts using high-speed technology, these ski resorts also had the highest vertical drops. In other words, the initial adoption decision was in many cases motivated by the same reason for the ski resort's continued persistence in adoption of faster lift capacity making the year of initial adoption endogenous.

Rather than an epidemic informational flow, ski resorts that attract primarily avid and vacation skiers engaged in endogenous competition in quality over time leading to an expansion of lift capacity and a higher proportion of faster lifts. At the national level the main determinant of differences in the degree of persistence was the ski resort's vertical drop. On the other hand, while a ski resort catering primarily to local skiers was more likely to make initial adoptions if other ski resorts had not done so, these ski resorts were unlikely to increase their proportion of faster lifts over time. In addition, ski resorts competing primarily in the national market were not constrained by skiable acreage, while the availability of skiable acres was an important limiting factor in the degree of persistence for primarily local ski resorts. Making sharp distinctions between local and national ski resorts is not without difficulty, but our results suggest persistence in adoption consistent with our expectations. More importantly, given the wide array of theoretical models that are of relevance to persistent adoption of innovations, we argue that there is value in knowing how an innovation affects an adopting firm's production process.

ENDNOTES

[1] There is also no role for consumer demographics, since all consumers are assumed to be identical.

[2] Mulligan and Llinares report that the cost of a detachable ski lift remained essentially unchanged in terms of nominal prices over a twenty-year period. On the other hand, there was little change in the relative price between it and the slower fixed-grip alternative during that time.

[3] Between 1978 and 1980 the number of items carrying bar codes increased from approximately 66 to 92 percent primarily due to the scanner's increasing ability to read smaller bar codes.

[4] Competitors' models were similar to those of IBM with competition coming mainly in the form of price reductions.

[5] Colombo and Mosconi (1995) provide an exception by finding no effect for this variable.

[6] For example, Hannan and McDowell; Levin, Levin, and Meisel (1987); and Karshenas and Stoneman (1993) have different results.

[7] For example, see Mansfield (1963), Hannan and McDowell, and Levin, Levin, and Meisel.

[8] The manager of the Pasadena Ralphs' store was quoted in 1980 as saying, "With scanners, we've been able to cut down from 12 checkouts to nine while still getting customers through at least 20 percent faster than before. And, as any chain-store executive will tell you, customers view the length of the lines second only to price when they decide where to shop" (Data Processor, 1980).

[9] This information was provided in a private communication by Mr. Randy Woolwine, Marketing Director for Dopplemayer-USA, one of the major suppliers of detachable chairlifts.

[10] The nominal price of a detachable chairlift has remained essentially unchanged at approximately $2.2 million per mile since 1981, while the current cost for the slower fixed grip chairlift of equivalent capacity (in terms of passengers transported per hour) is approximately $1.4 million per mile. In recent years the nominal price of the fixed grip chairlift has increased somewhat.

[11] In a personal correspondence, Mr. Woolwine, indicated that "very short detachable lifts are still quite expensive compared to fixed grip lifts. Reason: there is a lot of money in the terminals and associated safety systems. The longer the lift, the less the terminal cost is a percentage of the entire lift price".

[12]Narrowing the size of the market further, however, results in too many ski resorts without a local competitor. For example, in 1996 66 percent of ski resorts had no competitors located within 25 miles of the ski resort, while only 11 percent had no competitors within 50 miles.

[13] For example, in 1980 individuals in the top fifth of household income received 41.1 percent of aggregate income. By 1996 this figure had increased to 46.8 percent without accounting for the effects on wealth of increases in retirement and investment portfolios (U.S. Census Bureau, 1998).

[14]One exception was made for the ski resorts located near Denver, Colorado. In this case we defined the market as all ski resorts located within 125 miles of Denver.

REFERENCES

Allison, P. Survival Analysis Using The SAS System: A Practical Guide, SAS Publication, 2001.

Athey, S. and Schmutzler, A. "Investment and Market Dominance," *Rand J. Econ.* 32(2001), 1 – 26.

Barro, R. and Romer, P. "Ski Lift Pricing With Application to Labor Markets," *Amer. Econ. Rev.* 77 (1987), 875-890.

Bartoloni, E. and Baussola, M. "The Determinants of Technology Adoption in Italian Manufacturing Industries," *Rev. Ind. Org.* 19 (2001), 305-328.

Battisti, G. and Stoneman, P. "The Intra-firm Diffusion of New Process Technologies," *Int. J. Ind. Org.* 23 (2005), 1 – 22.

Cabral, R.and Leiblein, M. "Adoption of a Process Innovation with Learning-By-Doing: Evidence from the Semiconductor Industry," *J. Ind. Econ.* 49 (2001), 269 - 280.

Colombo, M. and Mosconi, R. "Complementarity and Cumulative Learning Effects in the Early Diffusion of Multiple Technologies," *J. Ind. Econ.* 43 (1995), 13-48.

Coyle, J. "Scanning Lights up a Dark World for Grocers," *Fortune*, (March 27, 1978), 76 - 79.

Cravatta, M. "Slope and slide (decline in popularity of alpine skiing)," *Amer. Demo.* 19 (1997), 34.

Das, N. and Mulligan, J. "Vintage Effects and the Diffusion of Time-saving Technological Innovations: the Adoption of Optical Scanners by U.S. Supermarkets," University of Delaware Working Paper 04-06, 2004.

Davidson, C. "An Equilibrium in Servicing Industries: An Economic Application of Queuing Theory," *J. Bus.* 61 (1988), 347 - 368.

"End of the Line," *Data Proc.* (June-July 1980), 7 - 9.

Ellickson, P. "Supermarkets as a Natural Oligopoly," Department of Economics, Duke University Working Paper, 2003.

Farzin, Y. Huisman, K. and Kort, P., "Optimal Timing of Technology Adoption," *J. Econ. Dyn. Cont.* 22 (1998), 779-799.

Fistel, L. "New Configurations Hone Efficiency in Retail Automation," *Can. Datasys.* 14 (June 1982), 35 - 39.

Fudenberg, D. and Tirole, J. "Preemption and Rent Equalization in the Adoption of a New Technology," *Rev. Econ. Stud.* 52 (1985), 383-401.

Geroski, P. "Models of Technology Diffusion," *Res. Pol.* 29 (2000), 603 - 625.

Giovanetti, E. "Perpetual Leapfrogging in Bertrand Duopoly," *Int. Econ. Rev.* 42 (2001), 671 – 696.

Goel, R. and Rich, D. "On the Adoption of New Technologies," *Appl. Econ.* 29 (1997), 513-518.

Götz, G. "Monopolistic Competition and the Diffusion of New Technology," *Rand J. Econ.* 30 (1999), 679 - 693.

Hannan, T. and McDowell, J. "Rival Precedence and the Dynamics of Technology Adoption: an Empirical Analysis," *Economica* 54 (1987), 155-171.

Kapur, S. "Technological Diffusion with Social Learning," *J. Ind. Econ.*, 43 (1995), 173 – 195.

Karp, L.and Lee, I. "Learning-by-doing and the Choice of Technology: the Role of Patience," *J. Econ. Th.*100 (2001), 73-92.

Karshenas, M.and Stoneman P. "Rank, Stock, Order, and Epidemic Effects in the Diffusion of New Process Technologies: An Empirical Model," *Rand J. Econ.* 24 (1993), 503 - 528.

Karshenas, M. and Stoneman, P."Technology Diffusion," In *The Economics of Technological Diffusion*, ed. P. Stoneman, Massachusetts, Blackwell Publishing, 2002.

Klepper, P. "Innovation over the Product Cycle," *Amer. Econ. Rev.* 86(1996), 562-583.

Levin, S. Levin, S. and Meisel, J. "A Dynamic Analysis of the Adoption of a New Technology: The Case of Optical Scanners," *Rev. Econ. Stat.* 69 (1987), 12 - 17.

Levin, S. Levin, S. and Meisel, J. "Market Structure, Uncertainty and Intrafirm Diffusion: The Case of Optical Scanners in Grocery Stores," *Rev. Econ. Stat.* 74 (1992), 345-350.

Metcalfe, J. "The Diffusion of Innovations: an Interpretive Survey," in *Technical Change and Economic Theory* (Dosi et al., eds), Printer Publishers, 1988, 560 – 589.

Mulligan, J. "The Economies of Massed Reserves," *Amer. Econ. Rev.* 73 (1983), 725-34.

Mulligan, J. "Technical Change and Scale Economies Given Stochastic Demand and Production," *Int. J. Ind. Org.* 4 (1986), 189-201.

Mulligan, J. and Llinares, E. "Market Segmentation and the Diffusion of Quality-Enhancing Innovations: The Case of Downhill Skiing," *Rev. Econ. Stat.* 85 (2003), 493 - 501.

Quirmbach, H. "The Diffusion of a New Technology and the Market for Innovations," *Rand J. Econ.* 17 (1986), 33-47.

Reinganum, J. "Market Structure and the Diffusion of New Technology," *Bell J. Econ.* 12 (1981), 618 - 624.

Reinganum, J. "Technology Adoption under Imperfect Information," *Bell J. Econ.* 14 (1983), 57-69.

Sarkar, J. "Technological Diffusion: Alternative Theories and Historical Evidence," J. Econ. Surv. 12 (1998), 131 – 176.

Stoneman, P. Handbook of the Economies of Innovation and Technological Change, Oxford, Basil Blackwell, 1995.

"Supermarket Scanners Get Smarter," *Bus. Week* (August 17, 1981), 88 - 92.

Sutton, J. *Sunk Costs and market Structure*, Cambridge, MA, MIT Press, 1991.

Sutton, J. *Technology and Market Structure*, Cambridge, MA, MIT Press, 1998.

Thomas, L. "Adoption Order of New Technologies in Evolving Markets," *J. Econ. Beh. Org.* 38 (1999), 453-482.

U.S. Census Bureau, The Official Statistics: Statistical Abstract of the United States, 1998.

Weiss, P. "Adoption of Product and Process Innovations in Differentiated Markets: The Impact of Competition," *Rev. Ind. Org.* 23 (2003), 301-314.

Chapter 7

TOWARDS AN EVOLUTIONARY THEORY OF PERSISTENCE IN INNOVATION

Christian Le Bas, *University of Lyon 2*
William Latham, *University of Delaware*

1.   INTRODUCTION

In this chapter we present a first coherent attempt to elaborate an evolutionary theory of persistence in innovation. We constrain our analytical framework in two ways. First, we do not make any attempt to explain why firms do not innovate. We recognize that a more comprehensive theory would explain a firm's decision to innovate on any basis (persistently, sporadically, or occasionally) or not to innovate at all. However, our objective is more limited in scope: we focus our attention on only innovating firms, seeking to understand why firms that innovate choose to do so persistently, sporadically or occasionally. These behaviors jointly constitute the phenomenon of persistence. Explaining persistence, not innovation *per se*, is our goal.

The second way in which we constrain our analysis is in terms of the sectoral technological trajectory as discussed in the in the Introduction to this volume. We are concerned only with sectoral technological trajectories along which firms invest in knowledge-producing activities such as R&D, design, and engineering. We do not attempt to account for the way in which persistent innovators carry out their projects in sectors where innovation is purchased from capital-goods producers. Nilotpal Das and James G. Mulligan, in Chapter 6, provide insights into this form of innovation.

Our discussion is organized as follows: In the first section we identify the fundamental concepts or principal themes that we have derived from evolutionary principles. Building upon these, in the second section we present a non-formal analysis of innovation persistence. In the final section we suggest a preliminary formal model which reflects aspects of the evolutionary tradition. It produces some interesting insights.

2.   PRINCIPAL OF THEMES OF AN EVOLUTIONARYTHEORY OF TECHNICAL CHANGE

Nelson and Winter's (1982) seminal analysis provides a base from which to construct new insights with the tools provided by subsequent evolutionary modeling. We begin with a brief review of salient parts of Nelson and Winter's work

2.1 INNOVATION PERSISTENCE IN A MODEL OF SCHUMPETERIAN INDUSTRY EVOLUTION: THE NELSON AND WINTER APPROACH

Nelson and Winter[1] analyze the complex interactive relationships among market structure, Schumpeterian competition and technical change. They emphasize firm behavior in terms of technological activity. Andersen presents a succinct view of their methodological presupposition:

> By accepting such . . . elements in their model-makers' tool-kit, Nelson and Winter are imposing upon themselves a certain conception of the evolutionary process. First of all, they apply a population perspective. An "industry" or an "economy" is seen as a taxonomic class incorporating a certain degree of variety of processes and/or products; but the variants must, in principle, be transferable between the different firms. This also implies a certain similarity of the search spaces of the firms, although there may be major differences with respect to the "distance" to different sources of knowledge...... Second, the name of the game is variety-creation and variety-selection within a given economic pattern." E. S. Andersen (1994, p 101)

In this context the basic assumptions they make are the following: In the industry under observation all firms produce the same identical product. This assumption this excludes the possible existence of "monopolistic competition" due to product differentiation. The industry output is defined as the sum of individual firm outputs. The product demand-price function is known and given, so that price is determined by industry output. The model envisages investment in R&D activity as simply generating higher productivity levels (there are no increasing returns due to firm scale). Firms can pursue two different policies: innovation and imitation. In this context imitation is not the same as replication.[2] Imitation may be a costly strategy, possibly even more costly than innovation. If patents provide effective protection of innovations, imitation of an innovation may require a product different enough from the innovating product for it to be successfully patented. Firm R&D

policies are assumed to be constant over time. The justification for such an assumption may be developed as follows:

> If competition is aggressive enough and profitability differences among policies are large enough, differential firm growth will soon make better policies dominate the scene, regardless of whether individuals adjust or not " (Nelson and Winter, 1982: 286).

The number of firms is given at the outset of the analysis. Entry of new firms is barred. The realization of an innovation (imitation) is an independent random variable. However the probability of success is proportional to the amount of R&D expenditures invested per unit of time. So the expected result of innovation (imitation) is considered to be a random draw. The productivity level is related to a random variable determined by a second draw. So, contrary to Dasgupta and Stiglitz's (1980) model, the firm's R&D policy is not ultimately derived from any maximization behavior. The Nelson and Winter (1982) approach defines a stochastic dynamic system in which, over time, productivity levels tend to rise and unit production costs tend to fall as better technologies are found and implemented

> as a result of these dynamic forces, price tends to fall and industry output tends to rise over time. Relatively profitable firms expand and unprofitable ones contract, and those that do innovative R&D may thrive or decline. In turn, their fate influences the flow of innovations (Nelson and Winter, 1982; 287)[3].

From different simulations of the model it appears clear that there is a trend affecting concentration in the industry under observation: there is a real tendency for industry structure to show growing concentration over the period of the simulations. This phenomenon is, in fact, explained by the basic mechanisms that are crucial in the context of innovation persistence.[4] In effect the successful innovators receive supranormal profits. To the extent that growth is keyed to economic performance, successful innovators grow more that their competitors:

> if a firm is a successful innovator frequently enough or if one of its innovations is dominant enough, the consequences of successfully innovation may be a highly concentrated industry structure" (Nelson and Winter, 1982; 308).[5]

Such a trend is also empirically verified by Mansfield (1962). He noted the *persistence* of the advantage of successful innovators who grow faster than others, although this advantage declines over time. However, Mansfield did not explain why the growth rate differential between the two types of firms is persistent rather than transient.[6] We presume that in the Nelson and Winter (1982) simulation model the effect of growth (and profitability) on innovation leadership goes through capital accumulation which determines the volume of R&D expenditures.

The main drawbacks of their study rest on the absence of clear analytical schemes showing the relationships among profitability, growth, R&D investments and innovative performance.[7] But they provide useful insights for exploring the interactive process of dynamic competition, firm growth and industry concentration[8]. With respect to the mechanisms underpinning innovation persistence, their study of industry evolution seems potentially fruitful. From the point of view of this volume, we note that in Nelson and Winter's model innovation persistence is possible, plausible or even highly probable. At the core of their analysis is this mechanism: the direct advantages of successful innovators, in terms of profitability and growth,are transformed in the next period of time to in advantages in terms of amounts of R&D effort and in terms of successful innovations. We will retain this aspect of their analysis in the following developments.

2.2 THE MECHANISMS OF TECHNOLOGICAL CREATION, SELECTION AND ECONOMIC DYNAMICS IN EVOLUTIONARY ECONOMICS

Our goal in this section is to enrich the seminal Nelson and Winter approach with respect to innovation persistence analysis. Evolutionary economics identifies three tasks for itself (Dosi 2000, Metcalfe 1993, Saviotti 1996): a) to isolate the economic mechanisms which generate variety in behavior, b) to identify selection mechanisms and their properties which eliminate behaviors that produce results falling below a certain level of performance, and c) to identify mechanisms, such as increasing returns and endogenous innovation, which provide feedback from the process of selection to the generation of variety. Accomplishing these three tasks provides the agenda we wish to develop here. Because firm performance is necessarily at the core of our topic, we need to define precisely what we mean by performance. We follow Metcalfe and Gibbons's (1986) fascinating analysis and distinguish three dimensions of firm competitive performance: (1) efficiency, (2) fitness, and (3) creativity.

1) **Efficiency** is the ability of the firm to transform technological success (innovation) into economic success (profit).[9] It is measured by the firm's technological and commercial performance (the level of labor productivity for instance). Technological and commercial performance is an indicator of economic performance (profits). In fact

evolutionary economics argues that the firm must realize a level of profit equal to or higher than the level required by competitive selection. Otherwise it could not resist adverse selection pressures. Efficiency is simply a capacity to resist adverse selection. In this respect many authors have noted that successful innovation requires additional elements including:

- Effectiveness in transforming knowledge about technologies into knowledge about products (von Tunzelman, 1995). This is necessary, but not always sufficient, to transform technological success into commercial success.[10]

- Complementary assets (such as patents) for protecting the rents stemming from innovation (Teece, 1986).

According to our definition, an efficient firm is able to master the technological, commercial, and economic competencies required to produce an innovation that responds to market needs.

2) **Fitness** is the ability and willingness of the firm to transform profits into new capital and growth. It is both the capacity to invest in physical capital in order to increase the firm's productive capacity and the capacity to invest in intangible (especially intellectual) capital to increase the firm's productivity. This second dimension is measured, in Metcalfe and Gibbons (1986) analysis, as the ratio of the firm's growth rate of productive capacity to its growth rate of profit per unit of output. In short, fitness describes the capacity to invest. We also include in fitness the investment made by the firm in its R&D activities.

3) **Creativity** refers to the ability of the firm to enhance and improve products and processes or, to put it simply, to innovate. Creativity is the ability to perform research successfully, to transform the capital of knowledge into new technological industrial competence. In many aspects, knowledge, competence and learning are directly involved here (Dosi, 2000). A firm's technological performance and, to a certain extent, its creativity are underpinned by its knowledge and its "dynamic routines" in R&D activities in particular (Nelson and Winter, 1982) and core competencies (Metcalfe and Gibbons, 1986). Firm creativity is a function of the firm's capacity to manage intellectual capital. At a very general level we find here crucial management of trade-offs at the core of R&D activity (among others: absorption versus internal production, exploration versus exploration, inward looking versus outward looking).

These three dimensions of performance correspond to three capacities: a capacity to produce a certain level of profitability, a capacity to invest it in new capital goods and knowledge activities, and a capacity to create new technological and industrial knowledge. Firms necessarily differ in each of these three dimensions of performance.

This approach in terms of competitive performance is very general. It is closely related to an understanding of the working of an economic organization under the influence of selection pressures. To make the process more specific, we also need a clear conception of the sources and drivers of innovation. Dosi (1997) finds five to be particularly important:

-    *Technological opportunities* account for variations among industrial sectors in both the ease and the cost of producing (possibly persistent) innovations. Many studies have found that the greater the degree to which technologies are subject to frequent scientific discoveries, the larger are the opportunities for innovation.

-    *Incentives* to exploit opportunities depend crucially on the structure of the market including the strength of competition and the ease of entry of new firms. We agree with Dosi (1997) that incentives themselves are probably less important than competitive pressures in explaining the decision to invest in knowledge-producing activities.

-    *Capabilities* of the economic organization (firm) as the agent to achieve technological change. A capability that is key to successful innovation is to have a strong knowledge base, including both an R&D capacity and a well-trained workforce (Archibugi and Lundvall, 2001).

-    *Organizational arrangements and mechanisms* through which technological advances are searched for and implemented. Teece (1992) has shown that strategic alliances, constellations of bilateral agreements between firms and networking strategies are increasingly necessary to support sustainable innovative activities. Among the important factors is the firm's own organizational structure (the literature has identified, among others, M-forms versus U-forms, and J-firm versus A-firm). In addition the institutional background of the industry and the economy as a whole have inluence.

-    *Appropriability conditions* (including patentability of innovations) are crucial for generating and maintaining the rents stemming from leadership in technological activities (see Levin et al., 1987 and Cohen et al., 2000). Appropriability conditions directly affect efficiency as the capacity to transform technological effectiveness in economic performance. These conditions differ significantly among industrial sectors.

These five sources or drivers of innovation, albeit interrelated, are linked to the three dimensions of firm competitive performance: efficiency, fitness and creativity. Figure 1 displays how these relationships can work. Incentives essentially affect the decision to invest in knowledge activities. Technological opportunities codetermine (with firm creativity) the effectiveness of knowledge activities. Firm capabilities and the organizational arrangements jointly operate on creativity and efficiency. Finally appropriability conditions influence efficiency.

We assume that there is no entry into the industry during the period of time under observation. In other words, innovation is only provided by incumbent firms that achieve productivity changes incrementally through the accumulation of innovations. So we are mainly concerned with a routinized technological regime (Winter, 1984). Although large firms are likely to prevail, there may be a fringe of small- and- medium-sized firms active in innovation.

3. AN EVOLUTIONARY MODEL OF FIRM INNOVATIVE PERSISTENCE

An evolutionary model of economic change (evolution) must explain the joint dynamics of learning (how the economic agents learn and achieve new technological artifacts) and selection (how the environment selects good learning and technologies). We suggest that a relevant model of firm innovative persistence must articulate explicitly persistence in innovative activity and persistence in profitability above a certain level (which is near the average level for the industry). The latter is a condition of the former. In so doing we link firm persistence in innovation with profitability and growth of the firm at the core of the analysis[11]. Figure 1 displays the main mechanisms at stake. It indicates the existence of a loop at the firm level: economic performance[12] $\Rightarrow$ investment in knowledge activities $\Rightarrow$ innovation $\Rightarrow$ economic performance. The five drivers of innovation support this loop by operating positively at certain stages. As the key factors that push (or pull) innovation persistence, they maintain the process of sustainable innovation on a positive trajectory. If not (if they do not act positively) they break the loop and the firm cannot innovate persistently. In the following paragraphs we consider the matter in more detail.

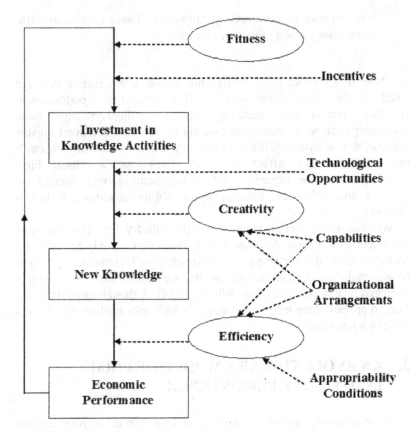

Figure 1 : The determinants of firm competitive performance

The message from Nelson and Winter's approach to industry evolution is that Schumpeterian competition produces winners and losers[13]. Some firms track emerging technological opportunities with greater success than other firms; "the former tend to prosper and grow, the latter so suffer losses and decline" (Nelson and Winter, 1982; 325). Success also confers an additional very important advantage: it makes further success more likely[14]. This outlines a dynamic process of competition and growth in which the three dimensions of performance (creativity, economic performance and fitness) are strongly interrelated.

Consider first a firm in a "virtuous" cycle defined by the following relationships: the firm has a high level of technological performance that it can easily transform into high or growing market shares. Therefore its level of profits is high, larger than the standard level of profitability for firms in the industry. Not only will the firm survive but it will get the means to improve its competitive advantage by investing a part of its growing resources in R&D activities (fitness) if the incentives it receive are high enough (we assume they are). If the firm's R&D teams are creative (due to astute management of its capabilities and organizational arrangements), the firm will innovate. One period of time later we will again find that the firm has a high level of technological performance, completing the cycle shown in Figure 1. In the next period another revolution of the cycle will begin, and so on. This intertemporal interaction between success in innovation, market shares, profitability and creativity has previously been modeled by Iosso (1993). It describes particularly well a dynamic process of persistence in innovation at the firm level[15]. In the same vein, Mansfield (1986) has identified one particular figure in this dynamic process, "the firm leader will normally remain the leader" over a long period of time.[16] It seems probable that among these leaders, there are many "first movers" because they achieve dynamic learning and produce persistent high levels of profit (see Mueller, 1997).[17]

Despite the seemingly irreversible reinforcing process described so far, the virtuous cycle, we also note that innovation persistence as an economic and technological process is not infinite. Eventually a time period may arrive when a firm which has previously been innovating persistently fails to remain in the ranks of innovating organizations. The model described here explains the factors that are in play. Any of the five drivers of innovation can be responsible for the cessation of innovation. Often it is the conditions and the directions of technological opportunities that change. In this case the firm may not modify its capabilities or organizational arrangements rapidly enough.

Now consider the case of a firm in "vicious cycle" which works along the following lines. The weak technological performance of the firm entails low and decreasing market shares. The inadequate level of profitability does not permit firm fitness. The level of investments in R&D activities is low or decreasing (in this case incentives are not at stake). There is a risk of too weak a level of technological performance (no or not enough innovation). In this configuration management of capabilities or judicious organizational arrangements could not offset the low level of investment in knowledge activity. This loop fits the process of technological sterility particularly well. The different mechanisms of this chain are not linked in a simple deterministic manner, there is place for firm strategy (good or bad), for chance in creativity and so on. Nevertheless the model predicts that, in general, such a firm will probably be eliminated from the economic system rather quickly. Recently Woicehyn and Daellenbach (2005) have assumed the existence of the same kind of process for industries that buy their technologies form outside. They argue that some firms possess a greater ability to integrate technologies into their operations than others. The factors that explain this greater efficiency for some firms are: strong strategic commitments to the technologies and their knowledge systems. In Woicehyn and Daellenbach's view there is a dynamic interplay of adoption process and knowledge system. They note that there exists a virtuous cycle for their firms just as we have described it here for firms that create their own new technologies.

To this point we have explained how the firm's efficiency, fitness and creativity can contribute to a high growth rate. Its rate of profit will be superior to the industry average, and the number of innovations will be high as well. We have now also explained the opposite case, that of the occasional innovator who does not succeed in surviving as an innovator (and probably not even as an economic agent, depending on the toughness of the competitive environment). But all firms do not follow one of these two evolutions. We have seen from empirical studies that firms also innovate sporadically. It is time to test if the model we suggest might also explain this type of firm dynamic.

Fortunately the same evolutionary model in which technological efficiency, fitness and creativity interact enables us to understand why firms innovate sporadically. There are two points at which mismatches can occur and give rise to breaks in the loop. Suppose a firm has limited efficiency due to irrelevant investments in marketing or errors in the management of complementary assets. Its level of profitability will be low, but sufficient to permit it to invest in new capital goods to increase its productive capacity and, most importantly, to invest a small amount in knowledge activities (R&D). It may be that this small volume of R&D is lower than the norm for the firm's industry, that is, the firm can pursue fewer R&D projects than the number required for innovating normally and regularly. But, and this point is crucial, if the firm has creative researchers and inventors, then its creativity can remain high from period to period. From time to time, this high level of creativity will offset the small number of R&D projects undertaken due to low efficiency, and thus the firm will innovate only from time to time. In this way the firm will innovate irregularly or sporadically, depending upon when its creativity offsets its low R&D. This phenomenon is made more likely by the fact that the product of R&D activities is, by its nature, irregular and uncertain. In this context the sporadic innovative behavior of the firm is explained by its limited efficiency which is sometimes offset by a high level of creativity (possibly due to effective management of its intellectual capital).

Now consider the situation of the firm whose average level of efficiency is higher than average in the long term, but irregular or cyclic. The amount of R&D projects undertaken will follow an irregular or cyclic time path as well. Suppose the firm's technological creativity is weak so that it does not produce innovations continually. When the firm's level of profitability is high, its level of R&D investment may be high enough to offset weak creativity and bring success in innovating. Conversely when the level of R&D activity is not high (due to weak profits), the firm's creativity is not strong enough to counterbalance the low profitability. Knowledge accumulation is not enough rapid. In this case the sporadic character of innovative behavior is mainly rooted in the irregularity of firm profitability.

Thus, in our analysis we can explain the sporadic character of innovative behavior by two different processes:

- a firm's low efficiency can sometimes be offset by very effective creativity[18] which may be determined partly by effective management of capabilities and partly by good choices in organizational arrangements,

- a firm's weakness in creativity can sometimes be offset by high efficiency.

The two processes by which firms innovate sporadically could be explained either by a low level of profitability followed by a lack of resources invested in innovation activity (low fitness) or by a deficiency in creativity. When the two failings are combined, the firm's trajectory converges outside the focus of this analysis since the firm no longer innovates and will have much more difficulties for surviving as an economic organization. These two explanations of sporadic innovative behavior might result in two different policy implications. The first focuses on the need to improve efficiency and the second focuses on the need to increase creativity.

In our analysis, there are two ways in which firm size may be important. First there may be a threshold size for persistent funding of innovation activity (R&D). Evidence of such a threshold has been found in R&D surveys.[19] Second there may be a size effect in creativity.[19] A larger firm has the resources to engage in more R&D projects at the same time, and with a larger portfolio of projects, it is more likely that at least one will be successful. On the other hand, smaller firms will be able to carry out fewer projects (perhaps none or only one) with a consequent reduction in the probability of success. Larger and smaller firms do not face equal levels of risk. Although innovation always entails some degree of risk, we expect firms to try to reduce or avoid risk. Large firms are able to do so by taking advantage of a "portfolio effect" for R&D projects.[20] In general smaller firms are less able to do this. So the risk of failing in creativity is stronger in the case of a small firm and this, in turn, interrupts persistence in the process of innovation. The result is that smaller firms are be expected to innovate less frequently and large firms to do so with a greater frequency.[21]

From this analysis we identify two important relationships: the relationship between profit and investment in innovation activity (R&D)

on the one hand, and the relationship between investment in innovation activity and capacity to produce innovations (creativity) on the other. We do not, as yet, have the data needed to investigate the first relationship between profit and investment in innovation activity.[22] More has been done on the second relationship (between investment in innovation activity and capacity to produce innovations). Understanding the mechanisms by which firms acquire, store and develop their technological, organizational and economic competencies provides one means of understanding their rates of change in technological performance and creativity. Teece and Pisano (1994) argue that the winners in the global market have been firms that can demonstrate timely responsiveness and rapid product innovation, coupled with the management capability to coordinate and redeploy internal and external competencies. Their competitive advantage has its roots in their "dynamic capabilities". This term emphasizes the role of strategic management in adapting, integrating, and re-configuring internal and external organizational skills and competencies in a shifting environment. We have shown in Chapter 2 that "dynamic capabilities" constitutes the key variable in explaining how the firm can continue to innovate persistently. Finally our analysis explains how firms accumulate knowledge: they accumulate profits from their technological performance (efficiency), invest in knowledge activities, produce new knowledge and use it productively. If these two types of accumulation (profit and knowledge) work well, the firm will likely enter the virtuous cycle of innovation persistence. If not, but one dynamic is working well enough to offset (under some conditions) the inadequacy of the second, a sporadic path in innovative behavior results. We have also shown that these two dynamics are different, but necessarily interrelated. The notion of "dynamic capabilities" further supports this view. Our model also shows that a persistent innovator is a persistent investor in R&D activity as many empirical studies have noted. But the reverse is not always true. The relationship between R&D investment and innovation performance is not totally deterministic but stochastic (Loury, 1979; Lee and Wilde, 1980; Nelson and Winter, 1982). It may be that the level of creativity is not sufficient to ensure the transformation of R&D into innovation.

It is also possible to use our model for different technological trajectories. In science-based industries the relationships between average success from R&D efforts and the subsequent technological advances are likely to be different from those in a cumulative-technology industry.[23] In the framework of our model it seems that, in science-based industries, for example, the recognition and mastering of technological opportunities is of paramount importance, while in cumulative-technology industries the management of creativity and choices in terms of organizational arrangements are more crucial.

Finally we have considered an evolutionary process involving the selection of "superior" or relatively more efficient firms over a long period of time. The distinguishing characteristics of the surviving firms are their fitness to grow and their creative capacity to discover new technological artifacts[24]. Our model of firm innovation persistence now encompasses all of the partial explanations presented in Chapter 1 in terms of technological accumulation and dynamic competencies.

4. A TENTATIVE DYNAMIC MODEL FOR FIRM INNOVATIVE PERSISTENCE

In this section we analyze implications of our evolutionary view of firm innovative persistence using a model which has only two simple equations. The first explains how innovation "turns knowledge into money". More accurately, Equation 1 describes how innovations explain profitability, and indirectly, the growth of the firm. Let I_{jt} represent firm j's level of innovation or the number of innovations produced in time period t and let P_{jt} be firm j's level of profitability in t. We propose that

(1) $\quad P_{jt} = f_j(I_{jt})$, with $f_j'(.)>0$

This relationship links two variables that are contemporary. Firm j's level of innovation, I_{jt}, can be measured as firm j's proportion of all innovations in the overall economy. Firm j's profitability can be measured as its rate of profit on all capital. Using these measures the two variables in the equation (1) are dimensionless numbers. Equation 1 is a model explaining efficiency.

We assume here that fitness is fixed and constant for all the firms in the economy. Consequently the levels of fitness and creativity might be put into the same equation that would show how the amount of investment in knowledge activity (R&D) explains technological performance, I_{jt}. this second equation explains how "research turns money into knowledge". Equation (2) summarizes a reverse relationship between innovation and profitability.

(2) $\quad I_{jt} = g_j(P_{jt})$, with $g_j'(.)>0$

Since firm fitness is fixed, Equation 2 describes only creativity. Equation 2 is related to the well-known knowledge production function used in many applied studies (see, in particular Griliches, 1995). It measures the cost of generating new knowledge or, alternatively, the productivity or creativity of the innovation process. It is what von Tunzelman (1995) has termed "the supply of innovation." It is the outcome of "technological accumulation" due to growth of the firm's knowledge through learning, either in the form of learning by doing or formal scientific learning, which reduces the cost of knowledge production. In Equation 2 we expect a rising monotonic relationship between *I* and *P* (the sign of the first derivative is positive). We suggest two possibilities regarding the rate of change in this monotonic relationship (i.e., the sign of the second derivative): (1) When learning is at a high level, the sign of the second derivative will be positive.(2) The converse is also true: when learning is weak, the sign of the second derivative will be negative. In terms of empirical evidence, the first case could indicate that there are dynamic economies of scale or learning effects in the production of innovation (Geroski *et al.* 1997). It seems equally possible to interpret the two options as describing two opposite situations as far as technological opportunities are concerned.

It is also possible to put a lag between the two variables of Equation 2. This would mean that, instead of current productivity explaining the current level of innovation, it would explain innovation in the next period:

(2a) $I_{i,t+1} = g_i(P_{it})$, with g_i' (.)>01

Combining the two equations (1) and (2a), we would have the reduced form:

(3) $I_{i,t+1} = g_i (f_i (I_{it}))$

Malerba *et al.* (1997) estimate an equation with this autoregressive form. Introduction of the lag also introduces dynamic adjustments into the model and permits us to examine the stability of innovation equilibria under alternative learning conditions.

As a first approximation, for illustrative purposes, assume that equation (1) is linear and rising monotonically:

(1a) $P_{jt} = \alpha_j + \beta_j *I_{jt}$, with f_j' (.)=β_j>0

Figure 2 illustrates the case of weak learning in the production of innovation and, conversely, Figure 3, shows the case of strong learning[25]. The two figures provide interesting insights. For each case we obtain two equilibrium points (*X* and *Y* in Figure 2, *X'* and *Y'* in Figure 3). At both *X* in Figure 2 and at *X'* in Figure 3 the steady state growth of the firm is lower than at *Y* in Figure 2 or at *Y'* in Figure 3 with a low rate of profitability and less innovation. The dynamic stability of the two points differs dramatically in the two figures. In Figure 2, *Y* is globally stable whereas *X* is unstable. This means that if the firm does not produce enough innovation, *i.e.,* less than I(*X*), its profitability will not be high enough for surviving as an innovating firm. In comparison, Figure 3 gives us another story: *Y'* is unstable whereas *X'* is stable. This means that a firm with a low initial level of innovation, *i.e.,* less than I(*X'*), will be able to survive as an innovating firm. Many studies (e.g., Geroski *et al.* 1997, Le Bas *et al.* 2003, and Cefis 2003) have found that there is a minimum threshold for inducing a long innovative persistence period of time for the firm. Here, our model indicates that this minimum threshold is not given once and for all. It depends on the type of learning in the production of innovation. When the learning effect is high, this minimum threshold seems not to exist. A firm with a very low level of innovation should converge towards point *X'* and stay persistently innovative. This point contradicts the usual results. Another difference occurs when learning effects are high: the equilibrium growth path (point *X'* in Figure 3) is lower than that which occurs when a low level of learning applies (point **Y** in Figure 2).

Our modeling, of course has many limitations. The primary one is that our view, based on a Marshallian "representative" firm, cannot take into account the co-evolution of the firm with its environment which is so important for shaping the innovative dynamics at the firm and sectoral levels (see the models in Dosi 2000). In the simple model we have presented here there is no entry, and no exit, and none of the type of turbulence which is related to the Schumpeterian process of creative destruction.

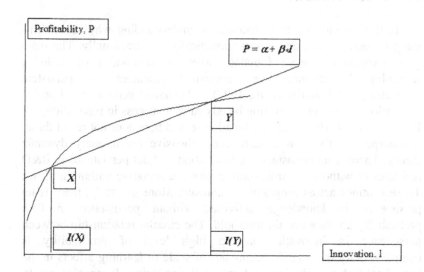

Figure 2. Illustration of Equilibria with Weak Learning in Innovative Activities

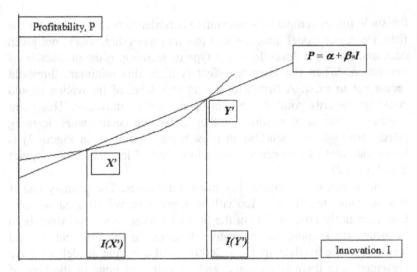

Figure 3. Illustration of Equilibria with Strong Learning in Innovative Activities

5. CONCLUSION

In this chapter we have focused on understanding why some firms innovate persistently, and others sporadically or occasionally. The main prior explanations were founded either on economies of scale in technological activities (*e.g.,* consistent patentees are persistent innovators, see Malerba *et al.*, 1997) or "dynamic economies of scale" (*i.e.,* prior success in innovation breeds current success in innovation, see P. Geroski *et al.* 1997). The explanation set out here does not reject these; it incorporates them in a more comprehensive evolutionary dynamic theory. There is no persistence in innovation without persistence in R&D activities or without dynamic routines that are operative within the firms. These routines act as programs to innovate. More generally, there is no persistence in knowledge activities without persistence in firm profitability above a certain threshold. The circular relationship between persistence in innovation and a high level of profitability is complementary to dynamic economies of scale or learning effects in the firm knowledge activities that we call creativity. It enables us to apprehend two different reasons for firms innovating sporadically. Finally we suggest a very preliminary formal model. Its essential prediction is that the minimum threshold size of innovative activity for inducing an innovative period for any firm is not given once and for all. The threshold size depends on the type of learning the firm experiences in the production of innovation.

Useful extensions of our model would take into account new firm entry (technological natality). This is very important in the entrepreneurial economy. Empirical studies (Geroski *et al.*, 1997, Le Bas *et al.*, 2003) have noted that many occasional innovators are small new firms. The entry of new innovating firms with high levels of technological creativity and industrial potential set up a powerful phenomenon that could eventually break the connections we have identified among efficiency, fitness and creativity. These new innovating firms may create a stronger competitive selection environment. For this reason it will be very important to continue to examine how technological natality (or technological entry) disrupts the virtuous cycle of persistence in innovation we have outlined here.[26]

ENDNOTES

[1] In particular their Chapter 12.

[2] "What distinguishes this situation from replication is the fact that the target routine is not in any substantial sense available as a template. When problems arise in the copy, it is not possible to resolve them by closer scrutiny of the original. This implies that the copy is, at best, likely to constitute a substantial mutation of the original, embodying different responses to a large number of the specific challenges posed by the overall production problem. However, the imitator is not directly concerned with creating a good likeness, but with achieving an economic success – preferably, an economic success at least equal to that of the original. Differences of detail that are economically of no great consequence are perfectly acceptable." Nelson and Winter (1982: 123).

[3] For a very clever analysis of the computational structure of their simulation model, see, among others, Endersen (1994).

[4] Nelson and Winter (1982: 308ff).

[5] There exist some factors that slow the trend. For example, concentration is smaller when firms invest more in imitation than in innovation for instance.

[6] "Growth confers advantages that make further success more likely." Nelson and Winter (1982: 325).

[7] In order to compensate for weaknesses of the Nelson and Winter model, some authors have extended the model in new directions. For example propensities to search innovatively and imitatively become variables in the Winter (1984) model, as does firm entry in a newly created industry. Product quality and product innovation are dealt with fruitfully in more recent studies (for a survey see Andersen, 1994). These improvements deserve more attention in the context of innovation persistence as well.

[8] We agree with Andersen (1994) that the Nelson and Winter modeling has helped others by demonstrating that it is not impossible to treat evolutionary economics processes in a systematic way; but they have not themselves provided a specification of an evolutionary economics research program.

[9] A writer for the Wall Street Journal Europe stated a similar idea thusly, "Research turns money into knowledge; innovation turns knowledge into money. " (Wall Street Journal 2004).

[10] For a review of this topic see Le Bas et al., 1998.

---

[11] Marris (1964) identified many factors that interact in the explanation of firm growth. Nevertheless, at the heart of firm behavior, he placed the firm's capacity to invest which, in turn, depends on its rate of profit. Penrose (1959) put the emphasis on firm diversification.

[12] As far as economic performance is concerned the reader will bear in mind that it is a relative or differential concept.

[13] Mansfield (1962) noted that, in terms of short-term growth, the rewards for successful innovators seemed to have been substantial, particularly for smaller firms.

[14] In Nelson and Winter's (1982) analysis, successful innovations are modeled by random draws, but the probability that a firm gets a successful draw is proportional to the firm's total innovation expenditures. In patent race models it is usually assumed that the research efforts increase the innovation production as well.

[15] Nevertheless there is a powerful phenomenon which can break this virtuous loop: the entry of new firms with higher creativity and technological innovations with high economic potential.

[16] In his study of the semiconductor industry Gruber (1994) confirms the existence of this sort of positive dynamic. Leaders do not all remain effective. There are obviously leaders that decline, see Bresnahan and Greenstein (1999).

[17] We have shown in Chapter 2 that a great majority of French innovative persistent firms over a long period of time are first-movers.

[18] If not, the firm will be an occasional innovator and will not survive as an innovator.

[19] We thank P. Saviotti for drawing our attention on this phenomenon.

[20] For example, Iosso (1993) shows that large incumbent firms achieve high productivity and profitability incrementally through the accumulation of many innovations and imitations. Firms with high output tend to perform much more research that might counteract any deficit in creativity. Current successes in innovation may flow from prior higher capacity.

[21] Although there is no entry in the model under consideration, it remains true that the main way in which small firms contribute to technological creation is through entry into the industry.

[22] Cefis (1999) has begun such analysis with a small sample of firms.

---

[23] See the simulations undertaken by Nelson and Winter (1982).

[24] It is clear for us that this evolutionary process does not necessarily lead to an optimal outcome. See Hodgson's remarks in Hodgson et al. (1994).

[25] In both figures we have shown cases of in which the parameters produce two intersections. Cases without intersections, with single intersections or with tangency solutions are considered less significant and are not examined here.

[26] Other empirical investigations have proceeded along similar lines. For example, Klepper and Simons (2005) have recently investigated the nature of firm survival patterns and industry shakeouts. Their study highlights the advantages of early entrants (first-movers) due to their proclivity to innovate. The early entrants become persistent innovators, achieving dominant market positions through innovation. Using a different line of reasoning in their empirical studies, Hatchuel et al. (2001) have shown that firms that innovate repeatedly may grow through long-term knowledge accumulation fueling the development of new products.

REFERENCES

Andersen E. S., (1994), Evolutionary Economics. Post-Schumepterian Contributions, Pinter Publishers.

Archibugi D., Lundvall B.-A., (eds.), (2001), *The Globalizing Learning Economy*, Oxford University Press.

Bresnahan T., Greenstein S., (1999), "Technological Competition and the Structure of the Computer Industry," *Journal of Industrial Economics*, March, vol. 47, pp. 1-40.

Cefis E., (1999), "Persistence in Profitability and in Innovative Activities," Paper presented at the European Meetings on Applied Evolutionary Economics, 7-9 June 1999, Grenoble.

Cefis E., (2003), "Is there persistence in innovative activities?" *International Journal of Industrial Organization*, vol. 21, n° 4, pp. 489-515.

Cohen W., Nelson R. R., Walsh J. P., (2000), "Protecting their intellectual assets: appropriability conditions and why US manufacturing firms patent (or not)," *WBER Working Paper Series*, n° 7552.

Daguspta P., Stiglitz J., (1980), "Industrial Structure and the Nature of Innovative Activity," *Economic Journal*, n° 90, pp. 266-293.

Dosi G., (1997), "Opportunities, Incentives and the Collective Patterns of Technological Change," *The Economic Journal*, vol. 107, (September), pp. 1530-1547.

DOSI G., (2000), *Innovation, Organization and Economic Dynamics*, Northhampton: MA Edward Elgar.

Geroski P., Van Reenen J., Walters C. F., (1997), "How persistently do firms innovate?" *Research Policy*, vol. 26, pp. 33-48.

Griliches Z., (1995), "R&D and productivity: Econometric Results and Measurement Issues," *in* : P. Stonemean, (ed.), *Handbook of the Economics of Innovation and Technological Change*, Blackwell, pp. 52-89.

Gruber H., (1994), *Learning and Strategic Product Innovation: Theory and Evidence for the semi-conductor industry*, Elsevier Science Publishers.

Hatchuel A., Lemasson P., Weil B., (2001), "From R&D to RID: Design Strategies and the management of innovation fields," Proceedings *of the 8th international product development management conference*, Entschedde the netherlands, EIASM.

Hodgson G., et al., (1994), The Elgar Companion to Institutional and Evolutionary Economics, Edward Elgar.

Iosso T. R., (1993), "Industry Evolution with a Sequence of Technologies and Heterogenous Ability. A Model of Creatrice Destruction," Journal of Economic Behavior and Organization, vol. 21, n° 2, pp. 109-129.

Klepper S., Simons K. L., (2005), "Industry shakeouts and technological change," International Journal of Industrial Organization, Vol. 23, n° 1-2, pp. 23-43.

Le Bas C., Cabagnols A., Gay C., (2003), "An Evolutionary View on Persistence in Innovation: An Empirical Application of Duration Model," in: P. Saviotti, (ed.), Applied Evolutionary Economics, MA: Edward Elgar.

Lee T., Wilde L. L., (1980), "Market structure and innovation," Quarterly Journal of Economics, vol. 94, n° 2, pp. 429-436.

Levin R. C., Klevorick A., Nelson R. R., Winter S., (1987), "Appropriating the Returns from Industrial Research and Development," Brookings Papers on Economic Activity, Special Issue on Microeconomics, vol. 1987, n° 3, pp. 783-831.

Loury G. C., (1979), "Market Structure and Innovation," Quarterly Journal of Economics, vol. 93, n° 3, pp. 395-410.

Malerba F., Orsenigo L., Peretto P., (1997), "Persistence of Innovative Activities, Sectoral Patterns of Innovation and International Technological Specialization," International Journal of Industrial Organization, vol. 15, n° 6, (October), pp. 801-826.

Mansfield E., (1962), "Entry, Gibrat's Law, Innovation and the Growth of the Firm," American Economic Review, vol. 52, n° 5, pp. 1023-1051.

Mansfield E., (1986), "Patents and Innovation: An Empirical Study," Management Science, vol. 32, n° 2, pp. 173-181.

Metcalfe J. S., (1993), "Some Lamarkian Themes in the Theory of Growth and Economic Selection," Revue Internationale de Systémique, vol. 7, n° 5, pp. 487-504.

Metcalfe J. S., Gibbons M., (1986), "Technological Variety and the Process of Competition," *Économie Appliquée*, vol. 39, n° 3, pp. 493-520.

Mueller D. C., (1997), "First-Mover Advantages and Path Dependence," *International Journal of Industrial Organization*, vol. 15, n° 6, pp. 827-850.

Nelson R. R., Winter S. G., (1982), *An Evolutionary Theory of Economic Change*, London: The Belknap Press of Harvard University Press.

Penrose E., (1959), *The Theory of the Growth of the Firm*, Oxford University Press.

Saviotti, P. (1996), *Technology Evolution, Variety and the Economy*, Cheltenham, UK and Brookfield, VT: Edward Elgar Publishing, Limited.

Teece D. J., (1986), "Profiting from technological innovation: implications for integration, collaboration, licensing and public policy," *Research Policy*, Vol. 15, pp. 285-305.

Teece D. J., (1992), "Competititon, Cooperation and Innovation. Organizational Arrangements for Regimes of Rapid Technological Progress," *Journal of Economic Behaviour and Organization*, vol. 18, n° 1, pp. 1-25.

Teece D. J., Pisano G., (1994), "The dynamic Capabilities of Firms: An Introduction," *Industrial and Corporate Change*, vol. 3, pp. 537-556.

Von Tunzelman, G. N. (1995), *Technology and Industrial Progress*. Cheltenham, UK and Brookfield, VT: Edward Elgar Publishing, Limited.

Winter S. G., (1984), "Schumpeterian competition in alternative technological regimes," *Journal of Economic Behavior and Organization*, vol. 5, no. 3-4, pp. 287-320.

Woicehyn J., Daellenbach U., (2005), "Integrative Capability and Technology Adoption," *Industrial and Corporate Change*, n° 2, (April), pp. 307-242

Chapter 8

PRINCIPAL FINDINGS, POLICY IMPLICATIONS AND RESEARCH AGENDA

Christian Le Bas, *University of Lyon 2*
William Latham, *University of Delaware*

1. DISCUSSION OF THE PRINCIPAL RESULTS AND NEW RESEARCH DIRECTIONS.

The Economics of Innovation emerged as an independent field of investigation at the beginning of the last century as has been described by Antonelli 2003, Dosi 1988, Dosi 2000, Rosenberg 1982, Stonemen 1995, and von Tunzelmann 1995. Then, very quickly, two specializations within, or approaches to, the subject emerged creating distinct subfields, each with its own theoretical road map. The first approach emphasizes the innovation itself, *i.e.*, the technology, and the second focuses on the innovator, usually a firm. Consequently, in the first approach, the innovation is the unit of analysis while, in the second, the firm is the central unit of investigation. In the first of these two approaches the analysis concerns how a new technology emerges, is adopted, developed, diffused, and incrementally improved over time. Technological trajectories, paradigm shifts (Dosi, 1982), technological systems (Carlson, 1997), and related concepts provide the basis for the models and schemes of analysis in this approach. The accent is placed on the accumulation of technological change along a trajectory or on the systemic interactions between technologies. Most innovation diffusion models, in particular those based on epidemic schemes, fall into this first category.

In the second approach the analytical focus is on the behavior of the firm *vis-à-vis* technological change. Strictly speaking there is not any behavior *per se* in the first approach: it presents a structuralist vision of technological change. When firm behaviors enter the scene, certain characteristics of the firms and their environments have been found to be particularly relevant in influencing the firms' decisions to invest in R&D or other knowledge-base-related activities (*e.g.*, design and engineering). These characteristics include the firms' own knowledge bases, incentive structures and technological opportunities as well as the firms' environmental characteristics such as market conditions and uncertainty. From knowledge of these characteristics it is possible construct an understanding of firm technological behavior, both in terms of conduct and performance.

The two approaches necessarily interplay more and less strongly along the type-of-analysis spectrum. It is a matter of observed fact that firms' activities with respect to innovation activity and technological improvements are endogenously determined. Firms shape their visions and determine their behaviors on the basis of the current state of the technology. For example, the evolutionary notion of technological regimes is typically, from this point of view, a hybrid concept in the sense it encompasses a technological dynamic together with some manifestations of Schumpeterian competition between firms. The industry life-cycle model is also a good candidate for a synthesis because it describes how industrial structures such as firm size, market concentration, and competitive interactions co-evolve with the rate and the direction of technological change. It seems to us that innovation persistence, as an economic and technological process, is crucially located at the intersection of these two dynamics. For example, the fruitful notion of industrial or technological *path-dependency,* which enables us to predict and explain the relative stability in the paths followed by technological change (and probably also by industrial change as well), might be analyzed as a by-product of the innovation persistence process. In effect, at the core of a path-dependent process we find increasing returns of some kind, positive feed-backs and, very often, irreversible lock-in. When a firm innovates persistently a path-dependent process is at work because the firm innovates in the vicinity of its own technological core-competence, along established trajectories that have been developed and improved. Technical change is produced locally and each improvement is localized (Antonelli, 1995). Technological paths correspond to a corridor along which firms innovate. Of course that does not imply that path-dependency would not exist if firm innovation persistence did not exist. The analysis simply implies that path-dependency would be much weaker without firm innovation persistence. Who can more effectively improve current technologies if not the knowledge workers of the firms that have conducted previous research on theses technologies? To some extent firm innovation persistence makes up some micro foundations for macro path-dependent trends.

In this book we have shown that there are firms that act as persistent innovators and have explained the likely mechanisms underlying this phenomenon. The idea of persistence in innovation activities has previously been questioned, even contested. For instance Aghion and Howitt (1992) doubt the assumption that the decision to increase investment in R&D activity today is explained by an increase in the expected volume of R&D that will be undertaken in the future. Aghion and Howitt propose that a negative relationship between present investment decisions and expected R&D can be derived as follows: When firms expect higher volumes of research by other firms, they are discouraged by the prospect of rapid technological obsolescence, which diminishes the expected returns to research. Conversely, when firms foresee low volumes of research by other firms, they are encouraged to engage in more research. Expected returns to research are increased because they believe that the next successful innovator is more likely to be able to retain a monopoly position over a longer period of time. This type of temporal interaction can produce a no-growth trap if many firms respond to expected high volumes of future R&D by curtailing their own present research. This view could be interpreted with the tools of the model set out in Chapter 7. It may be that there are insufficient incentives for repeated investment in knowledge activities.

The Aghion and Howitt assumption is only one among others; clearly it is not the only solution for understanding R&D conduct. Standard analysis, such as that of game theoretic models, has shown that under certain conditions, a monopoly threatened by a potential outsider (a contestable market) may innovate persistently. A reduction in the monopolist's perception of the level of threat will reduce persistence. A second explanation for reduced persistence is possible if there is a finite stock of potential technological advance. Then long-term innovative persistence tends to exhaust the stock of future technological opportunities. In this view, too much research today destroys the profitability of tomorrow's research. While there may be some validity to this proposition, we think that firms generally know how to manage the trade-off between exploiting established technological trajectories and exploring new ones. The problem is analytically similar to the optimal rate of exploitation of a growing natural resource (such as a forest or a fishery), well known in capital theory. A more significant challenge for firms with respect to innovative persistence is to learn how and when undertake a shift of trajectory.

In the following paragraphs we will discuss the main findings of the research presented in this volume in order to identify potentially fruitful directions for future research. To put it simply: knowing what we know, will help us to know what we do not know. In the model suggested in the Chapter 7 the basic idea is that present innovators draw some advantages from their past innovations. We have pointed out that these advantages can be marked. For example, when firms innovate and obtain early competitive advantages, they are then able to maintain or improve their economic performance, which enables them to fund additional R&D investments. Innovating firms are thus constantly initiating new cycles of research and innovation. In addition, past R&D (innovation) feeds a process of learning. Technological knowledge acquired in past is the main input for creating useful new technological artifacts. So, at the core of innovation persistence are two virtuous circles are at work:

(1) previous innovations provide the *means to invest* in preparing tomorrow's innovations, and

(2) previous innovations provide the *basic knowledge* through which new lines of innovation are possible (learning-by-researching). In Chapters 3 and 5 Alexandre Cabagnols has shown the significance of technological learning in the process of innovation persistence.

Of course the two virtuous circles, in which success breeds success and in which today's innovations push tomorrow's, are valid within the context of the exploitation of a previous technological trajectory (a minor innovation). The context of a trajectory or paradigmatic shift (a major innovation) is significantly different because it implies new competences that the firm cannot usually acquire through simple learning. In the research reported in this volume we have not explored innovation persistence in the context of trajectory or paradigmatic shifts. This subject deserves attention in a future research agenda.

It seems to us that what is going on when the firm experiences a process of innovation persistence in the context of a "supplier-dominated " sectoral trajectory is not so different. In Chapter 6 Nilotpal Das and James Mulligan present very striking results that contradict a thesis recently supported by Rosenberg: firms delay adoption in anticipation of a newer version appearing in the near future. Das and Mulligan observe that persistence in adoption continues across vintages even given a shortening of time between the release dates of subsequent vintages. The factor explaining this trend is the firm's incentives to differentiate the quality of its service from that of its competitors. This can be interpreted with the help of the virtuous cycle described previously: adoption $\Rightarrow$ better service quality $\Rightarrow$ differentiation $\Rightarrow$ growth of firm economic performance $\Rightarrow$ adoption of the newest vintage, etc. Moreover it is quite possible that a second virtuous cycle affecting the firm's stock of knowledge is occurring at the same time. But in this case the cycle is not based on learning in innovation (research) activities, since the firms in Das and Mulligan's theory do not produce innovations themselves but buy them the through the adoption of new equipment. The relationship is structured simply by a learning-by-doing process. Cabral and Leiblein (2001) have shown that the experience gained with the immediately preceding vintage increases the probability of adoption.

The analysis reported here highlights another advantage for innovating firms which provides an additional incentive for innovation. Innovating firms create barriers to entry based on market share, size and the ability to delay imitation (Antonelli, 2003). The effects of innovation persistence on entry deterrence have not been sufficiently explored, nor have the converse effects of firm entry on innovation persistence. Both of these deserve theoretical as well as empirical study and should be high on a new research agenda.

In Chapter 5 Alexandre Cabagnols finds that no specific country because effect is present when comparing innovative persistence dynamics at the firm level for France and the UK. This finding is counter to the literature on the importance of national systems of innovation. The lack of a country effect in the case of France and the UK might be because the two countries have similar characteristics. For example, they have similar amounts of patenting in the aggregate, similar industrial structures, etc. By contrast, what emerges from Chapters 2, 3, 5 and, to some extent, 6, is the existence of effects of the underlying technologies on the characteristics of the persistence path

Similarly the relationship between a firm's innovation persistence and the nature of its technological diversification, in the sense that the firm innovates in other technological fields more or less distant from the technological fields it has previously investigated, has not been adequately explored. This dimension may not have been investigated, in part, because technological diversification is now often studied in relation to firm internationalization. A second reason for the lack of study of the relationship between persistence and diversification may be because this type of shift causes structural change at the firm level, a phenomenon which makes analysis more challenging. Nevertheless this is a promising area for future research.

More generally the studies reported in this volume call for new analyses as well as clarifications regarding the macroeconomic consequences of innovative persistence. It would be useful and fruitful to first explore in theoretical terms some characteristics and general properties of aggregate economic dynamics that emerge from micro trend of innovation persistence. This could be undertaken in the same vein as the micro foundations of growth regimes modeled by Chiaromonte et al. (1993). One issue especially deserves careful study: are higher levels of innovation persistence likely to be correlated with more macroeconomic stability? To motivate the empirical analysis of this issue, we note, as reported in Chapter 2, that nearly 50 % of the overall number of patents granted to French firms are to persistent innovating firms. Surely this volume of persistent innovation affects macro productivity growth, but in which direction and in what proportion? In a first statistical treatment along these lines Le Bas and Négassi (2002) assessed the respective contributions of innovation persistence and technological natality to sectoral innovative performance. A future research agenda should attempt to address more precisely what the contribution of innovation persistence is to economic growth and stability at both the industry and the national levels.

2.   POLICY IMPLICATIONS

We believe that the research on innovation persistence reported in this volume has some important, though still preliminary, implications for public policy. There is now a consensus for acknowledging that, not only can individual firms innovate more, but also, more firms can innovate (Archibugi and Lundvall, 2001, p.12). An argument underpinning this book is an extension of this concept: not only can more firms innovate, but also, more can innovate persistently. We think, in particular, that more of the sporadic innovators that constitute a strong minority of firms as described in Chapter 2 could become persistent. Are there policies that would produce this result?

Here our concern is not that technology policy favor innovation activity in particular, nor is it with technology policy in general. Instead we would like to think about policy concerns which complement more closely the topics addressed in this book. Are there specific policy actions which could maintain or stimulate *persistence* in innovation activity. For many decades various forms of technology policy have been implemented with more and less success using such instruments as R&D subsidies, public procurement, implementation of technical standards, and competition policy (Metcalfe, 1995, Stoneman, 1995). But, to our knowledge, no studies have considered the outlines of a policy for persistence. It may be that one of the most significant results of this book for policy purposes will be wider recognition that the technology policies that promote innovation and the technology policy that enhance its persistence require different policy tools because the there are two different targets. In the chapter 3 Alexandre Cabagnols has shown very clearly that the crucial competences for innovation success (entry in the process of innovation) are mainly based on the establishment of cooperation and good management of external interfaces. By contrast, the set of competences at the heart of firm innovative persistence depend more on the development of in-house absorptive capacities (learning, training, and enhancing creativity). This view is in supported by our analysis in Chapter 7 of the virtuous cycle which affects knowledge through the interplay of firm growth and firm innovation persistence. It would be a potentially fruitful to analyze more completely the different kinds of entry into the process of innovation. There seems to be an obvious difference between a new idea that creates a whole new economic organization, a start-up, for example, and a new industrial idea (innovation) developed by an existing organization. Technology policies to promote these two kinds of entry into innovation may need to be very different.

In order to shed light on the main features of such a policy, we begin with Figure 1 in Chapter 7 which displays the factors and mechanisms involved in an evolutionary innovation persistence process. In particular we want to retain the five drivers which affect firm efficiency, fitness, and creativity: (1) technological opportunities, (2), incentives to innovate, (3) capabilities, (4), organizational arrangements, and (5) appropriability. While our discussion here will remain very general, we are aware that, because technology systems vary systematically among sectors and technologies, the implementation of a technology policy in practice requires detailed analysis of sectoral characteristics.

(1) Technological opportunities are a function of the state of technological knowledge. Opportunities are basic in the sense that they condition creativity. Fewer technological opportunities require larger R&D investments to maintain the level of innovativeness and, conversely more opportunities mean that less R&D investment can maintain the level of innovativeness. This observation implies that, to the extent that government funding of scientific and heavy, high-technology infrastructure can expand technological opportunities, persistent innovation will be encouraged by persistent government funding of such infrastructure. Recognition that the rapid growth and increasingly specialized nature of scientific and technological knowledge are creating technological opportunities is an essential ingredient of any innovation policy. From this point of view policies reinforcing technological opportunities are of crucial importance.

(2) Incentives to innovate are often considered as the core of an effective innovation policy. Incentives in this context frequently deal with market structures and with making the competition process more or less "aggressive." In our framework, and without underestimating this phenomenon, we adopt a view too often forgotten: demand has a strong influence on innovative activities, acting as powerful incentive for innovating. The idea that economic activity, *i.e.*, demand, pulls R&D or, more generally, innovative activities in a certain direction, goes back to the famous Schmookler (1966) study. There are now many variations in approach to the study of the way in which economic activities influence innovative activity. For example, P. Geroski and C. Walters (1995), examining the links between business cycles and patterns of innovative activity in the United Kingdom, concluded that innovations have a tendency to cluster during the boom periods. Their estimates provide clear evidence that causality runs from economic activity to innovative activity. Macroeconomic fluctuations also act as a powerful determinant of new firm start-up activity in periods of macroeconomic expansion and conversely. In a recent econometric exploration, Geroski et al. (2002) noted again that innovations are sensitive to variations in demand. Many others studies have noted the impact of the growth of demand as a variable explaining the persistence of innovating behavior (P. Geroski et al., 1997, Lhuillery 1996). In the same vein, Mansfield (1961, 1963, 1968) found that the speed of diffusion of innovations is significantly and positively affected by the rate of growth of demand. These studies confirm that economic macro policies that boost global demand could have a strong influence on the persistence of innovative activities (von Tunzelman, 2004). It is time to acknowledge the effects of macro policy on technology and on innovation persistence in particular.

(3) Firm capabilities. All policies that develop a firm's capacity to learn have to be considered as positive for innovation persistence. There is a huge and growing literature focusing on the "firm as a learning factory" and, more directly, on firm knowledge management and the market for technology Guilhon, 2001 both of which are certainly at the core of any persistent creativity strategy. By contrast, there are few remarks on the integration of innovation policy implications and schooling/learning policy. An exception is Von Tunzelmann:

> In order to maximize learning possibilities, it is crucial to take the labor force on board as part of the technological mission. The technology lifecycle progresses this may reverse itself, both through raising the level of learning among employees and through simplifying the access to technologies, but it may not do so entirely unaided". Von Tunzelmann (2004, p. 99)

At a very macroeconomic level the experience of some Asian countries bears witness to the reinforcing effect that schooling and learning have as an essential part of a long term technological strategy. Indeed this effect is a crucial part of the "miracle" of these countries' growth and technological advancement.

(4) Organizational arrangements. Recently Teece (1992) has brightly shown that new arrangements, such as the information of strategic alliances and the development of constellations of bilateral agreements among firms, are necessary for supporting innovation. They facilitate the complex coordination the price system can realize. To some extent they are an essential to the process of innovation persistence.

(5) Appropriability. This aspect is of paramount importance for transforming a technological success into an economic one. There are many means for protecting the rents stemming from innovation. We are aware that the causality in the relationship between persistent innovation and the means of appropriability runs in both directions. In some sectors persistent innovation is the usual mean for persistently maintaining an advantage over follower firms (because of production and distribution lead times and first mover advantages). As a result, in this particular context, persistent innovations are the lever of appropriability. In many others sectors patents and/or trade secrets are effective tools for protecting economic rents stemming from innovation. The optimum scope and duration of patent protection, *i.e.,* assuring sufficient protection to induce innovation without excessively slowing down knowledge dissemination, depend on the industry under consideration and the technology. Only good management of patents enables firms to protect their profits. National patent offices do not give an advantage to a sustainable inventor who applies for patent in a persistent manner. With this in mind we can examine the case of persistent innovation in

the form of a family of innovations stemming from the same first innovation. S. Scotchmer noted,

> There are no simple conclusions to draw about the optimal breadth of patents. It is not necessarily optimal to protect the first innovation so broadly that every derivative or second generation product infringes. If prior agreements are disallowed or not effective for some reason, then broad patent protection could discourage the development of second generation products,... and if the first innovator does not expect to profit by licensing to second generation innovators, broad protection could inhibit the first innovator as well thus undermining the entire research line. ... It appears that patent policy is a very blunt instrument trying to solve a very delicate problem. Its bluntness derives largely from the narrowness of what patent breadth can depend on, namely the realized values of the technologies. S. Scotchmer (1991)

To a large extent what emerges from this review on appropriability conditions is basically that this issue is a matter a firm management and strategy. Firms have to choose suitable complementary assets, relevant strategies as far as appropriability conditions are concerned, etc.

In summary this review seems to show that some drivers of innovation are more crucial for innovation persistence than others and thus that some policies will be more effective than others. For example, policies which expand technological opportunities will diffuse public knowledge for renewing firm knowledge bases, and policies which promote macroeconomic growth and stability as well as long term policies devoted to schooling and learning seem to be very effective. It is not certain that advocacy of these policies is compatible with the Schumpeterian view of economic development marked by crisis, cycles and creative destruction. But the basic message from this book is that an understanding of the roles played by creative and persistent accumulation of technological knowledge is also crucial to understanding economic development.

REFERENCES

Aghion, P. and Howitt, P. (1992), "A Model of Growth through Creative Destruct ion," *Econometrica* 60: 323-51.

Antonelli, C. (1995), *The Economics of Localized Technological Change and Industrial Dynamics*. Kluwer Academic Publishers. Boston.

Archibugi, D., Lundvall, B. eds., (2001), *The Globalizing Learning Economy*, Oxford University Press.

Carlson, B. (ed) (1997), *Technological Systems and Industrial Dynamics*. Kluwer Academic Publishers. Boston.

Chiaromonte et al. (1993). "Innovative Learning and Institutions in the Process of Development. On the Microfoundations of Growth Regimes," in Ross Thomson, ed., *Learning and Technological Change*. MacMillan Press, 117-49.

Dosi et al., eds., (1988), *Technical Change and Economic Theory*. Pinter Publishers.

Le Bas, C. and Négassi , S. (2002), "Les structures des activités d'innovation en France et comparaison avec celles des principaux partenaires commerciaux. Convention d'étude n° 19/2000 COMMISSARIAT GENERAL DU PLAN. Final Report November.

Le Bas, C., (2004), "Demand Growth as a Determinant of R&D Expenditures: An Empirical Model in the Schmooklerian Tradition," Revue d'Économie Industrielle, Mars.

Le Bas, C. and Patel, P. (2005), "Does internationalisation of technology determine technological knowledge diversification in large firms?" Revue d'Économie Industrielle

Efendioglu,U.D., von Tunzelman, N. (2001) Investment, Demand and Innovation: An empirical analysis for OECD countries, 1950-1995. Mimeo SPRU.

Geroski, P., Walters, C.F. (1995), Innovative Activity over the Business Cycle. *Economic Journal,* pp 916-928.

Geroski, P., Van Reenen, J., C. F. Walters (2002), "Innovations, Patents and Cash Flow," In Kleinknecht and Mohnen (eds) *Innovation and Firm Performance*. Palgrave.

Lhuillery, S., (1996), "L'innovation dans l'industrie manufacturière française: une revue des résultats de l'enquête communautaire sur l'innovation," in *Innovation, brevets et stratégies technologiques*, OCDE.

Lhuillery, S. (2003), "Les entreprises biotechnologies en France en 2001," Note de recherche. Ministère de la recherche. Septembre.

Mansfield, E.(1961), "Technical Change and the Rate of Imitation," *Econometrica*, 29, pp 741-66.

Mansfield, E.(1963), "The Speed of Response of Firms to New Techniques," *Quarterly Journal of Economics*. pp 290-311.

Mansfield, E.(1968), *Industrial Research and Technological Innovation*, Norton New-York.

Schmookler, J. (1966), *Invention and Economic Growth*. CUP. Cambridge.

Scotchmer, S., (1991), " Standing on the Shoulders of Giants: Cumulative Research and the Patent Law ," Journal of Economic Perspectives, Symposium on Intellectual Property Law, vol. 5, n° 1, pp. 29-41.

Von Tunzelmann, G.N. (2004), Integrating economic policy and technology policy in the EU.

INDEX

1. A. Phillips, A.P. Phillips and T.R. Phillips:
   *Biz Jets. Technology and Market Structure in
   the Corporate Jet Aircraft Industry.* 1994   ISBN 0-7923-2660-1
2. M.P. Feldman:
   The Geography of Innovation. 1994   ISBN 0-7923-2698-9
3. C. Antonelli:
   *The Economics of Localized Technological
   Change and Industrial Dynamics.* 1995   ISBN 0-7923-2910-4
4. G. Becher and S. Kuhlmann (eds.):
   *Evaluation of Technology Policy Programmes
   in Germany.* 1995   ISBN 0-7923-3115-X
5. B. Carlsson (ed.): *Technological Systems and Economic
   Performance: The Case of Factory Automation.* 1995   ISBN 0-7923-3512-0
6. G.E. Flueckiger: *Control, Information, and
   Technological Change.* 1995   ISBN 0-7923-3667-4
7. M. Teubal, D. Foray, M. Justman and E. Zuscovitch (eds.):
   *Technological Infrastructure Policy. An International
   Perspective.* 1996   ISBN 0-7923-3835-9
8. G. Eliasson:
   *Firm Objectives, Controls and Organization. The Use
   of Information and the Transfer of Knowledge within
   the Firm.* 1996   ISBN 0-7923-3870-7
9. X. Vence-Deza and J.S. Metcalfe (eds.):
   *Wealth from Diversity. Innovation, Structural Change and
   Finance for Regional Development in Europe.* 1996   ISBN 0-7923-4115-5
10. B. Carlsson (ed.):
    Technological Systems and Industrial Dynamics. 1997   ISBN 0-7923-9940-4
11. N.S. Vonortas:
    Cooperation in Research and Development. 1997   ISBN 0-7923-8042-8
12. P. Braunerhjelm and K. Ekholm (eds.):
    The Geography of Multinational Firms. 1998   ISBN 0-7923-8133-5
13. A. Varga:
    *University Research and Regional Innovation: A Spatial
    Econometric Analysis of Academic Technology Transfers.*
    1998   ISBN 0-7923-8248-X
14. J. de la Mothe and G. Paquet (eds.):
    Local and Regional Systems of Innovation, 1998   ISBN 0-7923-8287-0
15. D. Gerbarg (ed.):
    The Economics, Technology and Content of Digital T V,
    1999   ISBN 0-7923-8325-7
16. C. Edquist, L. Hommen and L. Tsipouri
    Public Technology Procurement and Innovation, 1999   ISBN 0-7923-8685-X
17. J. de la Mothe and G. Paquet (eds.):
    Information, Innovation and Impacts, 1999   ISBN 0-7923-8692-2

18. J. S. Metcalfe and I. Miles (eds.):
    Innovation Systems in the Service Economy:
    Measurement and Case Study Analysis, 2000   ISBN 0-7923-7730-3
19. R. Svensson:
    Success Strategies and Knowledge Transfer in
    Cross-Border Consulting Operations, 2000   ISBN 0-7923-7776-1
20. P. Braunerhjelm:
    Knowledge Capital and the "New Economy":
    Firm Size, Performance and Network Production , 2000   ISBN 0-7923-7801-6
21. J. de la Mothe and J. Niosi (eds.):
    The Economic and Social Dynamics of Biotechnology,
    2000   ISBN 0-7923-7922-5
22. B. Guilhon, (ed.):
    Technology and Markets for Knowledge:
    Knowledge Creation, Diffusion and Exchange within
    a Growing Economy, 2000   ISBN 0-7923-7202-6
23. M. Feldman and A. Link (eds.):
    Innovation Policy in the Knowledge-Based Economy,
    2001   ISBN 0-7923-7296-4
24. J. de la Mothe and D. Foray (eds.):
    Knowledge Management in the Innovation Process   ISBN 0-7923-7464-9
    2001
25. M. Feldman and N. Massard (eds.):
    Institutions and Systems in the Geography of Innovation,   ISBN: 0-7923-7614-5
    2002
26. B. Carlsson (ed.):
    Tehnological Systems in the Bio Industries:
    An International Study, 2002   ISBN: 0-7923-7633-1
27. D. Audretsch, R. Thurik, I. Verheul, and S. Wennekers
    (eds.): *Entrepreneurship: Determinants and Policy in a*
    European-US Comparison, 2002   ISBN: 0-7923-7685-4
28. J. de la Mothe and A. Link (eds.):
    Networks, Alliances and Partnerships in the Innovation
    Process   ISBN: 1-4020-7172-8
28. F. Belussi, G. Gottardi, and E. Rullani (eds.):
    The Technological Evolution of Industrial Districts   ISBN: 1-4020-7555-3
30. G. Fuchs and P. Shapira (eds.):
    Rethinking Regional Innovation and Change:
    Path Dependency or Regional Breakthrough   ISBN: 0-387-23001-7
31. W. Latham and C. Le Bas (eds.)
    The Economics of Persistent Innovation:
    An Evolutionary View   ISBN: 0-387-28872-4

SPRINGER